Getting It There

A Logistics Handbook for Relief and Development

by Ben Boyd

WORLD VISION INTERNATIONAL
919 West Huntington Drive, Monrovia, CA 91016

Getting It There
A Logistics Handbook for Relief and Development

by Ben Boyd

Published by MARC, a division of World Vision International.
Printed in the United States of America.

ISBN 0-912552-59-X

Table Of Contents

Foreword

Logistics formally became a part of World Vision International in 1984 when we began to participate in major emergency relief projects in drought-stricken Africa. Logistics has since played a significant part in each of our major relief projects.

It is satisfying to know that our organization now has the *Logistics Handbook* for our field teams to refer to when planning and implementing future relief and development projects. The magnitude, complexity and costs of our projects have become extensive; any miscalculation or misjudgment on the part of our logistics field teams not only could result in losses and wastes of donor contributions but could also significantly impact our relief ministry to depressed people. This handbook will serve as the basis for appropriate logistics policies and procedures to be promulgated by the International Office, and it will provide necessary direction for subsequent field projects.

It is my sincere desire that all of you who are assigned to WVI field project teams, especially those of you in key logistics positions, read and become thoroughly familiar with the contents of the *Logistics Handbook.*

Henry A. Barber III
Executive Vice President
World Vision International

Africa

Tunisia
Morocco
Algeria
Lybia
Egypt
Western
Sahara
Mauritania
Mali
Niger
Chad
Sudan
Senegal
Gambia
Guinea
Bissau
Guinea
Burkina Fasso
Sierra Leone
Liberia
Ivory Coast
Ghana
Togo
Benin
Nigeria
Cameroon
Central African
Republic
Djibouti
Ethiopia
Somalia
Equatorial Guinea
Gabon
Congo
Zaire
Uganda
Kenya
Rwanda
Burundi
Tanzania
Malawi
Angola
Zambia
Mozambique
Zimbabwe
Madagascar
Nambia
Botswana
South Africa
Swaziland
Lesotho

Preface

During the 1984-1986 crisis in Africa, World Vision became involved in several large-scale emergency relief projects. Because of the magnitude of these projects, with many of the commodities being furnished by support offices and other donors, it was necessary to formally introduce logistics into our organization. In many of these projects, thousands of tons of food commodities have been received, stored and distributed to hungry people by our field teams responsible for the accountability and ultimate disposition of these supplies. The total investment costs, including the costs of the commodities and resources, have been high. Consequently, in past relief projects, as well as future ones, there are no allowances for mistakes. The losses are too great.

The *Logistics Handbook* contains descriptions of the most common logistics functions that have been used in relief projects. To make these descriptions more vivid and to illustrate realities, I have included actual experiences encountered by our field teams. (These experiences are noted in italic print.) Some of the actions were appropriately taken; some were not. We trust that we can learn from these experiences--especially from our mistakes.

Although the handbook is centered mainly around our projects in Africa, it does not mean that these logistics activities are restricted to Africa. On the contrary, logistics applies to all geographical areas where our field ministries are active. This handbook, which has been designed as a reference manual, is expected to be used by field teams in all locations.

Even though this handbook has been directed mainly to the logistics team, it is understood that the project managers and field directors may not be fully experienced in logistics. These managers need to know what to expect from their logistics team and be able to evaluate their ability to get the job done. Chapter XV, the final chapter in the handbook, is directed to these managers. It provides a summary of the major logistics functions described in detail in previous chapters. The summary does not tell all, but it does give the managers insight on what to look for in these specific areas. It should be helpful to a manager who is trying to keep several balls in the air at one time!

The writing of this handbook has culminated in 29 years of my experience in the varied fields of logistics. It has been a privilege for me to have been a part of World Vision's logistics operations during the famine crisis in drought-stricken Africa. Many lives have been saved through our relief projects. My role has largely been to provide direct assistance and advice to the field teams. Finally, in the formulation of this handbook, I would like to give recognition to the following:

-- First, to World Vision for giving me the opportunity to travel extensively and become directly involved in the "hands on" logistics of their major relief projects.

-- To my brothers and sisters in Christ who are enduring the hardships, discomforts and loneliness by serving Him and World Vision in implementing the logistics plans in these remote regions.

-- To Dana Honn and her efficient team of word processers: Gayle, Renee, Stephanie and Jan, who did all the typing of the manuscript through several drafts. And, especially to Jacci Clark who spent many hours in proofing the manuscript.

-- To Cedric Taylor for his gifted talents in designing the cover and drawing other illustrations that appear throughout the book.

-- To Steve Heassler for his very special expertise of directing the laser typesetting and layout in preparation for the printing of the handbook.

-- To Kn and Davina Chow-Moy for their fine work at engineering the actual desktop computer production of the book.

-- To World Vision's management staff and field personnel who reviewed the manuscript and provided constructive comments.

-- To my wife, Elli, who gave me much encouragement during the agonizing and frustrating moments in the formulation and writing of the handbook.

-- And, finally to our Lord Jesus Christ who directed the entire project by providing His constant and never-ending love, wisdom and guidance to make it all possible.

When adversities occur in your project, constraints develop and the problems appear to be insurmountable, I encourage you to take your problems to Him in prayer. I can attest from my own personal experiences that through His amazing grace there is suddenly hope, and the situation soon becomes much brighter. Things just seem to fall into place. He seems to delight in wanting to help. All you have to do is ask Him and give Him the opportunity.

May God bless you and your project!

Ben Boyd
Logistics Specialist
March 1987

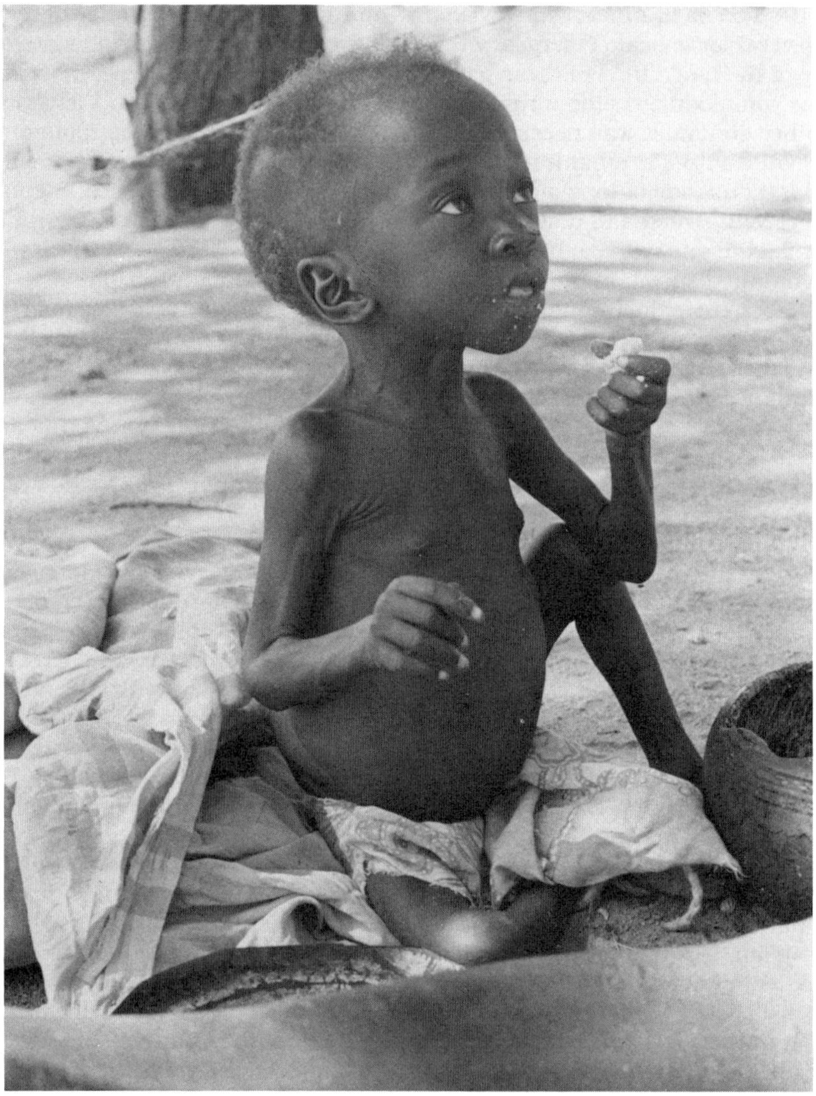

For He will deliver the needy who cry out,
the afflicted who have no one to help.
He will take pity on the weak and the needy
and save the needy from death.
He will rescue them from oppression
and violence, for precious is their blood
in His sight.

Psalm 72:12-14

I. Introduction

". . . for He satisfies the thirsty and fills the hungry with good things."

Psalm 107:9

The primary purpose of this *Logistics Handbook* is to provide procedural guidelines to World Vision field teams involved in planning and implementing relief and development projects. Because of the rapid turnover of field personnel, it is necessary to have effective continuity of a project. This handbook is designed to assist in this continuity, particularly as it pertains to the logistics support of a project. Other documents important to the continuity of a project are:

1. *The WVI Project Planning Workbook.*
2. *The WVI Project Proposal Form and Supplmentary Project Information Form.*
3. *Logistics Plan (sometimes referred to as Distribution Plan).*

Because of the importance of logistics planning in any field project, the objective in the preparation of the handbook has been to encourage planners to refer to these guidelines. It is recognized that any handbook of this type is of no use if it is not used by those who are directly responsible for the planning and implementation of the project. Consequently, the handbook is laid out in such a way that it can be used as a ready reference by planners as they become confronted with any particular logistics function.

What is Logistics?

The term "logistics" has in the past related to military operations. In fact, Webster's dictionary begins the definition as "The aspect of military science dealing with . . . " World Vision has applied its own definition as "The aspect of World Vision's support operations dealing with the procurement, transportation, materials management, warehousing, maintenance and other matters relating to the handling of the details of a relief or development project."

By necessity, logistics has become an essential part of all industrial and commercial operations requiring attention to the details of the common logistics functions addressed in this handbook. As World Vision becomes more involved in relief and development projects, particularly in the remote locations of Third World countries, logistics plays an ever-increasing role. Traditionally, logistics is somewhat of a mystery to those who have had little or no experience in the business. However, some overly confident individuals, viewing themselves as instant experts, may charge ahead without adequate forethought and detailed planning. When they get into trouble, they panic and call for the nearest logistics specialist to bail them out. But it is often too late--the damage has already been done.

There is nothing overly technical or highly sophisticated about logistics. Much of it is common sense and good judgment. But logistics matters do not take care of themselves. Someone must be designated to think logistics and plan for the logistics support of an operation. If the project manager does not have a logistics specialist on his staff, he should designate someone with experience to handle the logistics matters, or be prepared to handle them himself. As stated in the definition, logistics is handling the details of an operation. Since logistics is the product of an orderly thought process, a logistics plan must be a part of every relief and development operational plan.

(Refer to Page 15 and also Appendix V for instructions on preparing a logistics plan.)

Ingenuity in Sudan and Kampuchea

Ingenuity is very important to the successful execution of a logistics plan. A successful example of the use of ingenuity occurred in the Central Region of Sudan in 1985. Our truck distribution of food commodities to hungry villagers in the Blue Nile province was unexpectedly interrupted by early rains. The soil condition in this part of Sudan is quite different from the sandy soil that absorbs the water in other parts of the country. Here, because the soil becomes very sticky and gummy when it gets wet, the roads become totally impassable. Our team knew that there were many hungry people in and around the village of Kurmuk in the south. When all else failed, our team organized a camel caravan to deliver the bags of grain. The camels were able to trudge through the mud, each carrying about five 100-pound sacks of food. Because of this ingenious action and a deep-seated determination to get food to hungry people, many lives were spared. This is logistics!

Africa is not the first area in which World Vision has been in the logistics business. In 1980 in Kampuchea, long before the main thrust of our efforts towards the famine crisis in Africa, we became involved in a major logistics operation.

World Vision was one of the first relief agencies to be invited to return to Kampuchea after the Khmer Rouge had slaughtered nearly 3 million of its people over a span of four years. The country was on the brink of a mass famine. Thousands of children had been orphaned by the atrocities, and many who were suffering from malnutrition continued to roam the countryside in search of food. World Vision's Board of Directors authorized an expenditure of $11 million to be spent in the first year with an objective of bringing as much relief and comfort as possible to the remaining 5 million depressed people.

A preliminary survey was hurriedly made in Phnom Penh to identify significant projects where World Vision could become involved; the survey team was headed up by the president at that time, Dr. Stan Mooneyham. We gave assistance in the following areas: orphanages, medical and agricultural colleges, pig and poultry farms, vocational schools for teenagers, a nursery for orphaned babies, a large rice development project, a condensed milk factory, the Kampuchea Red Cross and the commercial fishing industry.

The one project that was of prime interest to both World Vision and to the newly formed government of Kampuchea was reinstating our

A camel caravan loads food at Damazin in the Central Region of Sudan to take to the village of Kurmuk. When there was an urgent need for the food, no other transport could reach this area at the time.

In early 1980 relief agencies found the few remaining survivors of Pol Pot's devastation on the brink of mass famine.

pediatric hospital in Phnom Penh. Four years earlier when construction had just been completed and the 60-bed hospital was ready to be opened, the Khmer Rouge moved in and literally gutted the building and destroyed all of the equipment. The building was used as one of the many execution chambers in Phnom Penh by the ruthless Khmer Rouge. Approval to rebuild the hospital was the subject of intense negotiations between the Prime Minister and Dr. Mooneyham. The government finally granted approval, and work began immediately. Eight months later, Dr. Mooneyham and the Kampuchea Minister of Health dedicated the newly renovated hospital. This pediatric hospital, under management control of World Vision, is the only hospital of its kind, providing care and treatment to the children of Kampuchea.

To provide the support for our team in Kampuchea, a logistics base was established in Singapore, managed by World Vision's field office. Hundreds of tons of building materials and other supplies were purchased in Singapore and flown in to the team in Phnom Penh. There was practically a full-time charter of a C-46 aircraft flown by a crew from the Philippines. This plane made at least three trips each week--often more. The carrying capacity of the C-46 was about 2-3 MT. Later, as emergency relief demands increased, World Vision acquired a four-engine cargo plane in

A chartered C-46 aircraft is unloaded in Phnom Penh, bringing relief supplies into Kampuchea.

The World Vision-owned Argosy aircraft delivers a load of pigs and chickens to Kampuchea in 1980 from our logistics base in Singapore.

A converted LST was chartered to bring supplies from the Philippines for a rice-planting project in Takeo Province of Kampuchea. In the background is all that is left of a bridge across the Mekong River that connected the city of Phnom Penh which the Khmer Rouge destroyed.

Australia called an Argosy. Both aircraft maintained a busy flight schedule hauling relief goods and supplies into Phnom Penh. The Argosy, with an Australian crew, was a very versatile aircraft with a payload of about 8 MT. On more than one occasion, the Argosy met the requirements of the pig and poultry farm project by flying in sows, chickens and ducks. All construction materials used in rebuilding the pediatric hospital, except roof tiles and sand for mixing concrete, were flown in from Singapore. This busy airlift lasted almost one year. These aircraft were the lifeline of our project.

One of our projects in Kampuchea included the complete support of a rice-planting project in Takeo Province, situated near the Vietnam border. Total requirements to cultivate some 15,000 hectares were provided, which included rice seed, fertilizers, insecticides, motor-driven plows and irrigation pumps. These requirements were marshalled and shipped from the Philippines on a chartered Landing Ship Tank (LST), a converted World War II vessel which was used in military amphibious operations. Since Phnom Penh was situated on the Mekong River, the ship was able to dock at the city pier where the cargo was unloaded onto leased trucks for direct delivery to Minister of Agriculture warehousing in Takeo.

This emergency relief project represented a major logistics operation. It could not have been accomplished without the bases of support provided by our teams in Singapore and the Philippines which handled all the procurement, contracts, marshalling, storage, accountability and shipping of all the vast tonnage levels of goods and supplies needed in Kampuchea. When the cargo arrived in Phnom Penh, the team arranged for direct deliveries to the respective projects where the supplies were closely monitored.

No two projects are exactly the same; thus the logistics functions will not always be the same. Flexibility is the key to a successful logistics operation. The logistics support must be fitted to the operational plan. Generally, any functional area relating to the support of the project will be assigned to the logistics section by the project manager. The logistics organization should be tailored according to the functions being performed. (Organization for Logistics in a project management team will be covered in Chapter III.)

Commercial Application of Logistics

The following logistics effort of the Trans-Alaska pipeline project is mentioned to provide World Vision's logistics specialists with a better appreciation of what logistics is all about and the importance of identifying early any logistics constraints. Some of your constraints may not be as obvious as in the following scenario, but they could be equally critical to your relief or development project.

During the planning phase of the Trans-Alaska crude oil pipeline, the construction management of Alyeska, a company formed by the seven major oil companies which funded this project, was neither prepared nor organized to meet logistics challenges. The eventual costs of this project were in excess of $9 billion. In terms of costs, personnel, equipment and supplies, the Trans-Alaska pipeline project was the largest single construction job ever undertaken in the USA by private companies. In the initial two years of planning, Alyeska used several engineering consulting firms to form an elaborate and detailed construction plan.

This project was unique in many ways. It offered challenges not only to the logisticians, but also to the construction engineers. One-half of the pipeline had to be built across virgin tundra under extreme weather conditions. After the construction plan was completed, it then became necessary to develop the construction support (logistics) plan. If there had been any appreciation for the logistics aspects of such a major project, both plans would have been developed at the same time.

Alyeska contracted Bechtel Corporation of San Francisco (one of the largest architectural and engineering construction firms in the USA) to provide construction management for the pipeline. Fluor Corporation provided construction management for the 12 pump stations along the 500-mile pipeline route. Neither Bechtel nor Fluor had a properly oriented logistics organization or logisticians when they began their initial planning. However, upon the insistence of Alyeska, Bechtel and Fluor began their search for experienced logisticians to add to their organizations. Not too surprisingly, many of the ones they hired were retired logisticians from the military services.

Before any pipeline or pump station construction

These two cable-controlled Hoverbarges were designed to transport construction supplies across the Yukon River in Alaska in 1975. This was a major logistics challenge during the construction of the Trans-Alaska pipeline.

Supply trucks wait to cross the Yukon River on the Hoverbarges.

could begin north of the Yukon River, it was necessary to build a haul road to the source of the crude oil--Prudhoe Bay. This haul road would generally follow the planned route of the pipeline. Ultimately, there were seven construction camps erected along this haul road. Each accommodated over 500 construction workers and provided supply, warehouse and maintenance facilities in support of the construction effort.

Before the entire project was completed, several major logistics and environmental constraints seriously affected the construction schedule. Each of these constraints represented challenges to the logistics staff. Problems of weather, terrain and environmental protection had to be overcome in order to resolve these problems.

One of the major logistics constraints identified early in the planning phase was the Yukon River--a major obstacle that would affect the haul road and later the pipeline construction. Crossing into the virgin country of northern Alaska, the Yukon is one and one-half miles wide. It flows at a constant rate of seven knots per hour, and in winter it is frozen over with ice of varying degrees of thickness.

To meet the completion schedule dictated by the sponsoring oil companies, there could be no delays or interruptions in year-round construction. Crossing the river was essential since it provided the only means to move heavy construction equipment, tons of supplies, steel pipe, vehicles, people and everything else needed to complete the project. A solution had to be found to breach the Yukon River obstacle.

The river-crossing problem was given to the logisticians with technical assistance from the engineering staff. They spent numerous hours in brain-storming sessions considering every conceivable means. The solution that seemed to be the most feasible, reliable and cost effective (and the one eventually used) involved a cable-controlled hoverbarge. The staff prepared contract specifications and sent them to the few companies which appeared to have the technical expertise to put together such a system. Nothing of its type and size had ever been used before. The first company to be awarded the contract failed because the hoverbarge it built was not capable of maneuvering across the treacherous ice formations.

The second contractor was a British firm which was the leader in that country for building hovercraft (air-cushioned vehicles). Components of this contractor's hoverbarge were flown from England and assembled on the banks of the Yukon. The huge hoverbarge was designed to carry up to 90 tons, thereby accommodating the heaviest and largest piece of equipment, or up to three truckloads of main pipe sections. The hoverbarge was a huge, flat platform with powerful engines driving large fans on each side of the platform.

There were rubber skirts all around the bottom edge of the platform to contain the air produced by the fans. The air would lift the barge sufficiently high enough to allow the barge to maneuver over the snow and ice formations. Then, with a huge wench on each river bank, the hoverbarge was pulled and guided across the river by a steel cable. This system proved so successful that a second hoverbarge was added.

Responsibilities for Logistics

In WVI, every effort is being made to assign professional logisticians to each major field team. The recruiting of experienced logistics specialists has been extremely difficult because of the shortage of qualified applicants who have strong Christian commitments. (Chaper XII addresses these logistics personnel problems.)

If there are no logistics specialists assigned to a team, the project manager should designate one of the other staff officers to handle logistics matters, or handle these matters himself. The logistics aspects of a project must not be ignored.

In 1984, soon after World Vision became involved in emergency relief projects in Africa and the first logistics specialist joined Africa Drought Project (ADP), it did not take too many trips to the field to realize a checklist of logistics actions would be useful to field teams planning relief projects. Such a checklist was hurriedly prepared and widely disseminated to all active field teams. This document was used quite extensively, and helped to carry us through the major relief projects with reasonable success. Mistakes were made. However, considering the vast tonnages of food commodities and supplies that have been handled by World Vision since early 1984, there have been relatively few losses resulting in any significant wastes of funds and resources.

In President Tom Houston's "ADP Evaluation" report of March 27, 1986, he had this to say about logistics:

"EVALUATION: It is ironic that this was not defined when it proved to be one of the most critical performance areas. We hired a professional logistician who provided working papers, staff visits and hands-on assistance to the ADP effort. In spite of our relative inexperience as an organization with a logistical challenge of this magnitude, we were able to procure goods, provide transportation, storage and distribution, though not without error. Given our lack of staff, our performance in this area was credible, and has earned us the esteem of many government and Non-Government Organizaton (NGO) leaders. We have been criticized as well, and are the wiser for our logistical efforts in the delivery of food and medicines."

Because logistics planning and execution is here to stay in World Vision relief and development projects, this handbook is a must. It provides the necessary continuity to compensate for the rapid turnover of field team members, and it also provides an important link with the International Office.

The logistics specialist is a staff officer who is closely involved in the operations, working directly with the project officer and project manager in planning and implementing relief and development projects. The logistician cannot do his job if he is not brought into the operational plan early in the project. Any logistics constraint must be identified early in order that appropriate actions can be taken to eliminate this constraint, or the operational plan must be altered accordingly.

Project Planning Workbook

The Project Planning Workbook has been designed to assist field teams in making a professional and effective disaster survey and project design. The workbook serves as a guide, checklist and planning aid in the development of a project. It serves as the basis for filling out the Project Proposal Form (PPF) and Supplementary Project Information Form.

Included in the workbook is a Logistics Planning section. An outline for a general logistics plan is provided, with a partial checklist as a guide.

The Project Planning Workbook should be completed by the field team when the project is first conceived. It will serve as an excellent guide in the preparation of the PPF. In fact, the preliminary logistics planning will identify early any major constraints. The essentials of the initial logistics planning should be included in the PPF to satisfy the International Office that the logistics aspects of the project have been considered. The project manager should require his team members to complete their appropriate section of the handbook. The logistics specialist is responsible for Section 10.0, Logistics Planning. The preparation of a more detailed logistics plan should follow immediately. (Refer to Appendix V for a sample outline of a logistics plan.)

Logistics Plan

There is no set format or procedure for the preparation of a logistics plan. World Vision field teams have not been in the practice of preparing a separate logistics plan in the development of a project, but this is important and should be done. A written plan provides the only assurance that all essential logistics aspects of a project are covered. In most cases, the project officer will cover certain logistics matters interspersed throughout the overall operational plan, PPF or distribution plan. The review of several past operational plans has shown that many essential logistics details have not been covered. These omissions in the planning phase have resulted in costly mistakes and miscalculations. A separate logistics plan, including the details of each functional area, ensures that all essential support functions will be planned for.

After determining what must be done in the operational project, whether it be relief or development, the logistics specialist (or whomever the project manager designates to handle the logistics operations) evaluates the project in view of the major logistics functions. He identifies those functions that will play a part in the operational project. There may be some questionable areas such as staff housing, air operations, radio communications, etc., where the project manager's decisions may be required--whether or not he desires to assign these functions to logistics. At this stage, the project manager understandably will have much on his mind. Consequently, the logistics specialist must be prepared to advise and make appropriate recommendations to the project manager. The logistics specialist is now ready to prepare an outline for his logistics plan.

Logistics is simply exercising common sense and good judgment. It requires the logistician to *THINK!* In a quiet atmosphere, you must visualize everything that must be done in handling the food commodities from the very beginning when World Vision takes possession of the food commodities until they are distributed and placed in the hands of the receiving family or village chief.

As an example of a practical application of another form of logistics, let's assume that, if you are married, you are planning a vacation trip with your wife and children. After you and your wife decide what you want to do on your vacation you, as the "logistics planner," begin to think of and visualize all the things you need to do to arrange for this trip. Since you will be going out in the "bush," if you forget anything it will not be convenient for you to return for it. Also, since there are so many details involved, you cannot trust your memory. You must sit down and make a detailed checklist. This is a list of things that may be easily referred to in order that nothing will be overlooked. You certainly do not want to arrive at your campsite and then realize you forgot the fuel for your cookstove or matches or flashlight or the baby's Pampers--all are essential to a successful camping trip.

This practical example is not fundamentally different from preparing a logistics plan. That is why it is important to prepare a written plan so that nothing will be overlooked. For your relief project, you can develop your own logistics checklist. Even if you are not married and not involved with baby's Pampers, this type of thought process would apply in a similar activity as it relates to your own personal planning.

In some of World Vision's relief projects in the drought countries of Africa, we have had logistics problems during the implementation phase, largely because there was no individual designated by the project manager to handle the logistics planning. If your staff has not been blessed with having an experienced logistics specialist assigned to it, at least designate another staff officer to be responsible for logistics and then call on the International Office or the regional office for assistance. Do not simply ignore these matters, or expect them to take care of themselves. It is the responsibility of the International Office to provide experienced logisticians who can assist you.

Do not be concerned about preparing an elaborate logistics plan. As the acting logistics specialist, you need to organize your thought process to make certain that all essential logistics details have been covered. Use the Project Planning Workbook to prepare a separate logistics appendix to the operational plan.

Shown in Appendix V is a sample outline of a logistics plan. By necessity, the sample logistics plan shown is hypothetical. It is an outline that includes all the major logistics functions. The outline contains a checklist of significant actions that should be considered when preparing the plan. Any particular function that would not be applicable to the project would be omitted.

As was mentioned, the logistics plan should be simple and concise, but sufficiently complete in detail to document exactly how each logistics function is to be performed. The written plan should be a recorded thought process of the visualization of how the project will be logistically supported. One important purpose of a written plan is to make certain that nothing in the area of logistics support is overlooked. When logistics oversights and omissions in planning occur, this is when the budget is exceeded and the project costs become greater than originally planned.

Now, since World Vision's field teams are in the learning process of logistics, training media such as this handbook will assist future teams in becoming more professional when preparing and planning projects.

It must be emphasized that logistics must be kept simple and remain flexible. Because of its name, some people have a tendency to avoid logistics, thinking that it is some "weird" technical science in which special training and education in the field are required. This is not the case. But it does require attention. In your project, *KEEP IT SIMPLE!* Don't make it complicated. Make absolutely certain, however, that you, if you are the project manager, assign someone on your staff to handle logistics matters. If not, you must do it yourself, and this is dangerous since you have many other things to think about. It is best to have a specific individual on your team who can *THINK LOGISTICS!*

II. Concept of Operations

". . . I tell you the truth, whatever you did for one of the least of these brothers of Mine, you did for Me."

Matthew 25:40

For World Vision relief and development projects, logistics actually begins at the distribution centers in the country that is being supported. What is a distribution center? It is a location selected by the World Vision project team where there are facilities for warehousing food. The distribution centers should be reasonably close to the villages, refugee camps or feeding centers which will be receiving the food commodities. They should be located near main roads accessible to large capacity long-haul trucks. They could also be near a railroad or railhead if there are storage facilities available and a convenient place for unloading.

It is normally the donor's responsibility to deliver the food commodities to the distribution centers designated by the World Vision field project team. At this point, the project team assumes full responsibility and accountability. The transporter who delivers the commodities to the centers is responsible for unloading the trucks or rail cars and placing the commodities in the warehouses at the direction of World Vision's warehouseman. After counting the containers (bags, boxes or drums) and noting the condition of the commodities, the warehouseman signs the appropriate bills of lading carried by the transporter.

(NOTE: Reference is made throughout the handbook that the responsibility for the control and accountability of food commodities begins at the distribution centers in the project area designated by the field project teams. It should be emphasized that this handbook is directed to the logistics specialists who are members of the project teams. Particularly when PL480 commodities from USAID are donated to World Vision, World Vision's responsibility should begin when the ships are loaded at the U.S. ports. However, this responsibility is normally assumed by the freight forwarder who acts as the World Vision agent to ensure that the commodities are properly delivered to the field distribution centers. At this point, World Vision personnel actually assume the responsibility in a hands-on role.)

The commodities will normally be delivered in 50 kilogram bags, equal to 110.25 pounds (kilogram x 2.205 = pounds). (For conversion factors, refer to Appendix IV.) In case the grain is obtained in other African or European countries, the bags could weigh as much as 100 kilograms. Because the handlers are laborers who are weak from hunger, these heavy bags can be extremely difficult to handle. If World Vision is able to specify the size and weight of the bags, they should be specified to not exceed 50 kilograms in weight.

Even though World Vision project teams are not responsible for the commodities until they arrive at the designated distribution centers, you may wonder how those commodities even got to these remote distribution centers. Assuming that the grain is being shipped from the point of origin by ocean freight, World Vision in coordination with the donor will obtain the services of a freight forwarder who will make all the necessary shipping arrangements, including truck or rail transport from the overseas port to the land-locked country and eventually to your distribution center. (Chapter VIII, "Ocean Freight," will explain more of these particulars and will include the duties and responsibilities of the freight forwarder.)

During the famine crisis in Africa, the majority of World Vision's emergency food was donated by the U.S. government. These commodities were largely surplus agriculture products, purchased from American farmers and shipped from the U.S. Gulf Coast ports. The donor which was the U.S. Agency for International Development (USAID) representing the U.S. government, arranged for the freight forwarder. There was coordination with our own World Vision Relief Organization (WVRO) in selecting the freight forwarder and scheduling the shipment. Other sources of food commodities used in the African relief projects were gifts-in-kind from other World Vision partners including

Canada, Great Britain, Australia and New Zealand.

Following are some examples of relief projects and the source of large tonnages of food commodities:

1984 (Mozambique)

World Vision was in partnership with the Dutch government which provided 10,000 MT of white maize purchased from the Malawi government and shipped by rail, air and truck to the Tete Province of neighboring Mozambique. World Vision's role in the partnership was to coordinate the transport, distribution and accountability of the white maize to the hungry people of Tete Province.

1985 (Mozambique)

USAID donated 17,000 MT of yellow maize to World Vision. Most of the maize was shipped by ocean freight from the U.S. Gulf Coast ports to Port Elizabeth in South Africa. This food was shipped in essentially five separate shipments, each about two to three months apart. At Port Elizabeth, the freight forwarder's representative arranged for the ships to be unloaded and the cargo to be placed on rail cars for further shipment to Harare, Zimbabwe. World Vision contracted a local company in Harare which met the arriving rail cars and transloaded onto trucks for movement to the contractor's warehouse in Harare. The same contractor later transported the bags of maize to designated warehouses and distribution points in Mozambique at World Vision's direction and supervision.

1985-86 (Sudan)

World Vision's introduction to Sudan in 1985 was in partnership with World Food Program (WFP), which was receiving allocations of food commodities from USAID. The area of responsibility, situated around the Blue Nile, included two provinces of the Central Region south of Khartoum. There were thousands of hungry people in this region, most of whom were displaced Sudanese in search of food and water. There were also many Chadian refugees.

WFP did not have the capability to distribute and monitor these commodities. From the beginning of July to the end of December 1985, World Vision project teams did a phenomenal job of distributing an average of 15,000 MT of food per month.

In FY86 World Vision established its own program with USAID for this same region, taking over from WFP. The total allocation was slightly in excess of 26,000 MT.

In addition, a gift-in-kind of over 3,000 MT from the Canadian government, arranged for by World Vision Canada, was received at the Port of Sudan and distributed in the Central Region. Most of these commodities were supplemental foods for use in the nutritional feeding centers. The ownership of these commodities passed to World Vision at Port Sudan. Out of the 26,000 MT, about 15,000 MT were to be purchased locally by USAID. In this case, ownership of the food, mostly dura (sorghum), passed to World Vision at the various government agricultural bank warehouses. The grain was either relocated to World Vision's leased warehouses at the two distribution centers in Wad

A Sudanese woman prepares a meal from food distributed from World Vision's warehouse in Wad Medani in the Central Region.

A Nomadic mother and child receive nourishment from a feeding center near Menaka in eastern Mali.

Families in the village of Mucumbura (Mozambique) eagerly wait to receive their allocation of maize from across the border of Zimbabwe.

Madani and Damazin, or distributed directly to the villages from the government warehouses. World Vision contracted truck transporters for both the long-haul from Port Sudan and the short-haul from the two distribution centers to the receiving villages.

1985-86 (Ethiopia)

Our emergency relief projects during the extreme famine crises in Ethiopia were the largest in terms of tonnage and the number of hungry people who were cared for. The concept of operations involved the passing of ownership of the food commodities at the Port of Assab. For the most part, World Vision owned and operated its own trucks in Ethiopia, both for long-haul and short-haul. World Vision bought and erected warehousing at the depots (distribution centers) located at Kombolcha and Nazareth.

1985-86 (Mali)

In early 1985 a World Vision team made a preliminary survey of the 7th Region of Mali. This is a very large area that comprises the extreme eastern part of Mali, east of the Niger River. The area is almost all desert. Stories of hunger and severe malnutrition among the children and adults of this area were confirmed by the survey team. It was found that the problems were mainly among the nomad tribes whose livestock had all died off because of the drought. This left the nomads with nothing to trade for grain and other foods, which had been their custom. The only other relief agency operating in the area was the International League of Red Cross. But because the area was so large and the hunger problems so great, it was not possible for one agency to be effective. World Vision saw an emergency situation in and around the desert village of Menaka, with many children dying each day from malnutrition and from an epidemic of measles.

When the relief machinery finally was in gear, USAID agreed to negotiate a barter arrangement with the country of Ghana, which had just experienced a surplus crop of maize. They needed rice and vegetable oil. There was not time to put a program together and ship food from the United States and expect it to arrive in Mali before the beginning of the rainy season, which usually starts in June. Obtaining maize from Ghana would save valuable time.

The barter arrangement was successful, and 8,000 MT was to go to three distribution centers in Gao, Menaka and Ansango all located in the 7th Region. Another 2,000 MT was to go to the 1st Region around the regional capital of Nioro where many villages were without food and were suffering from the drought conditions.

In 1985, the peak year of the famine crisis in Africa, it is estimated that a total of 9.7 million hungry people, including malnourished children, received emergency relief assistance from 667 different World Vision projects at a cost of some $44 million to World Vision. In the funding of these relief projects, World Vision was responsible for all storage and distribution costs of the commodities from the distribution centers to the villages, as well as the costs of personnel, administrative support, transport, maintenance, fuel, spare parts, etc. In most cases, the cost of inland transport required to deliver the commodities to the distribution centers was the responsibility of the donor.

It is important for project logistics teams to know that ownership and responsibility of the food commodities do not pass to field teams until these commodities reach the storage facilities at the designated distribution centers. The truck transporter, operating under the supervision of the freight forwarder or his representative, delivers the commodities to WVI warehouses and provides laborers to unload the trucks and stack the commodities at the direction of the WVI warehouseman. The official document that transfers the ownership is the bill of lading provided by each truck driver. The signature of the WVI warehouseman on this bill of lading relieves the truck driver of his responsibility for the commodities, and he is paid after returning the signed document. The warehouseman signs for the exact number of bags he receives and verifies the condition of the cargo.

In some cases, the drivers have been careless and have not adequately protected their cargo in transit. Bags of food have presumably been stolen in transit. Sometimes drivers have not protected the bags and several have become wet in transit. All bags must be accepted, but a careful count should be made of the bags received which are wet or potentially spoiled. These bags are to be isolated from the good bags and will become subject to a claim against the carrier.

The grain spoils as it dries and, if eaten, will make the people very sick. In Mali wet bags were retained and kept segregated from the good bags, to be used later for animal stock feed. This could be dangerous when the people are hungry; they will risk sickness when they can obtain food regardless of its condition. If the people become ill from eating World Vision-supplied food, any positive image or reputation World Vision might have had would be destroyed. It is the responsibility of the logistics team that this does not happen.

In cases of lost, stolen or damaged commodities, a formal claim is filed against the transporter for his negligence. These claims are filed, based on the warehouseman's inventory and the notation of the condition of cargo upon the truck's arrival, and the claims are generally the responsibility of the freight forwarder, or his representative.

WVI project teams are cautioned not to interfere with the truck transporter's operation or the freight forwarder before the cargo arrives at WVI storage locations. This part of the operation is not World Vision's responsibility.

During the Mali project, when maize was being transported by truck from Ghana to WVI-designated storage locations in the 7th Region, some project team members interfered before the trucks arrived at their destination. The interference occurred at various locations north of Niamey (Niger) when a WVI logistics team member was on separate resupply missions from Niamey in support of the nutritional feeding centers in and around Menaka.

The contractor who arranged for the truck haul was to ensure that the drivers would travel in convoy and be properly supervised while the trucks were in transit through other African countries. Because there was little or no control over the drivers, several of the trucks became lost, and some drivers were reluctant to travel the difficult roads into Mali. The WVI team member encountered some of these trucks along the road and, knowing that this was World Vision food which was needed very badly in Menaka, attempted to give the drivers instructions which they did not want to accept.

In another situation, the team member ordered unloading and disposition of the food before it reached its destination. The team member acted in good faith, but he exceeded his authority. The result could have been embarrassing for World Vision but, as it happened, no significant quantities of food were lost or spoiled.

The logistics team in the field should clearly understand exactly when and where the ownership of the food commodities becomes their responsibility. From this point on, the logistics team must ensure that adequate protection, storage and accountability are maintained over these food commodities. The control is World Vision's responsibility until the commodities are placed in the hands of the villagers or village chiefs. This responsibility is generally assigned to the logistics team. The steps and procedures for handling this control will be described in subsequent chapters of the handbook.

III. Organization

". . . Defend the cause of the weak and fatherless; maintain the rights of the poor and oppressed."

Psalm 82:3

Because of the flexible nature of logistics, there is no set logistics organization that would apply to all projects. Each project must be analyzed or evaluated in terms of the actual functions of logistics needed to carry out the operational plan.

Fill Key Positions

If your project runs true to form, as most World Vision field projects have run thus far, it is one thing to identify the applicable logistics functions, but it is another to be able to fill each key position in your logistics organization. It has usually been extremely difficult to find qualified personnel with the necessary skills, experience and Christian commitment to fill these positions. You may have to just do the best you can with the personnel you have. Once your key positions have been identified and approved at the regional and International Office level, appropriate standard personnel requisitions can be submitted through the Human Resources Department (HRD) channels for necessary recruiting action.

Whenever possible, attempt to find qualified persons locally. The logistics manager position should be filled by an experienced logistician. Like all World Vision field teams, the objective is to have a small nucleus of experienced expatriates to provide on-the-job training for local nationals who will one day become qualified to assume the responsibilities of the position. Because of the shortage of qualified personnel, you may have to combine certain functional positions. The logistics manager must ensure that the work load is equally distributed so that one or a few individuals on the logistics team do not become overloaded. Potentially qualified local nationals should be used fully, in order that they can be trained in the functional areas.

From experience, our problems in logistics support have been largely the result of a lack of organization and qualified people. Invariably, it has been common for us to undertake a project before adequate resources and qualified people are in place to do the job. This is understandable for our type of ministry. But in some cases, this method of doing business has created considerable hardship on the part of the few employees present and has raised some criticism by other PVOs and governmental organizations.

> *When an initial survey disclosed the starving people in Menaka in eastern Mali in early 1985, WVI was so eager to bring relief to these affected people that they started an operation long before they were prepared. In Mali many lives were saved long before the teams were adequately staffed and equipped, and for that we are all grateful. There was, however, great stress and strain on the few staff members who were there doing the job.*

It is difficult to criticize this method of operation. However, our administrative support of these teams and personnel recruiters must be more responsive to meet the needs of these teams in a more timely manner.

Avoid Spreading the Project Too Thin

Another area where we need to profit by our mistakes is to avoid too much fragmentation within country projects. Where limited resources are available, multiple projects should not be maintained when they are not able to be adequately supported.

> *Again, using Mali as an example: With the size of the country and its limited transportation and lack of communications, WVI established two separate projects situated in the extreme opposite ends of this huge landmass, over 1,000 kilometers apart. There was no possible way these two projects could be mutually supporting.*

In cases such as this, particularly where there are shortages of people, equipment and vehicles, consider every possibility of consolidating and selecting a geographical area that can be adequately supported. In countries where there has been such great need as in the drought-stricken countries of Africa, there is an enormous desire and tendency to want to do too much, more than the project team is capable of handling. It is better to do a job well that is manageable, than to do a job poorly that is beyond our capability. Start off small and gradually expand the operations as additional resources are made available to the project.

Ideally, before any project is initiated, it is best to have at least the key people in place, vehicles and other equipment on hand and the means to maintain them, warehousing built or leased, and any other essential support requirements in place. Granted this is not always possible, largely because of the emergency nature of the relief project. The situation should be different with development projects because there is more time to plan and get the necessary resources and personnel in place.

Use Good Judgment and Common Sense

Good judgment and common sense must be the rule when implementing relief projects before these essential resources are in place and ready to function. Supporting headquarters such as the field offices and the International Office must be prepared to respond in a timely manner to the needs of the project team, so they can effectively and efficiently accomplish their job. When these teams must struggle without the key positions filled, the necessary funding, and the necessary means available to do the job, the team morale will drop.

Model Organization

1. **Management**: Figure III.1 shows the logistics organization which was eventually established in World Vision Ethiopia in late 1985. This particular organization can be used as a model organization since it includes all the major logistics functions that you would expect to be represented in a major relief project.

   It will be a long time before World Vision will again participate in an emergency relief project which would even come close to matching the

The grim expressions on the faces of this Kampuchean father and his malnourished child show the hunger and strife that prevailed until the refugees arrived in the World Vision camp in Thailand.

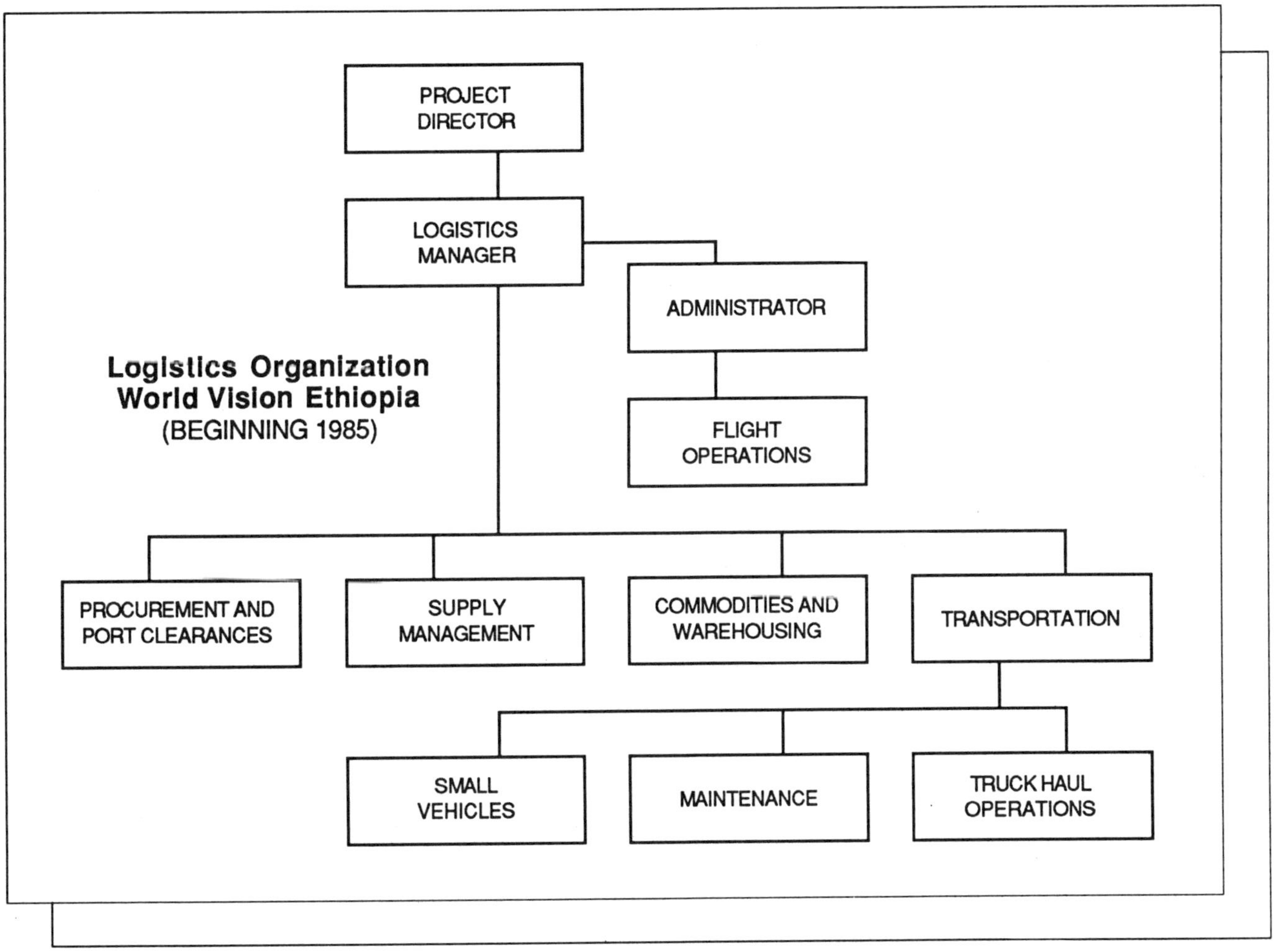

FIGURE III.1

project in Ethiopia in terms of staff, equipment, tonnage levels and costs. When the initial buildup began in late 1984, it was most unfortunate that World Vision did not ensure that an experienced professional logistician be assigned to the team. A more stable, experienced logistics team could have prevented many of the mistakes in organization and judgment that created numerous growing pains in this project.

During the peak period of the emergency relief operations in Ethiopia, the magnitude of the project and the tonnage of food commodities handled during this period created a monstrous logistics operation. World Vision was operating over 12 major feeding centers with monthly feedings in excess of 500,000 people. The project included both dry food distribution as well as nutritional wet feedings to malnourished children, adults, lactating mothers and the elderly from these major feeding centers. Most of the food commodities were furnished by the U.S. government, and the ownership of these commodities passed to World Vision at the Port of Assab. Because of the limited numbers of trucks available in the country and the demand placed on these trucks by the government of Ethiopia to support their military operations, World Vision was forced to purchase and operate its own truck fleet. This fleet consisted of both long-haul and short-haul trucks and trailers and will be discussed in more detail in Chapter VI, "Transportation."

The majority of the feeding centers were located to the north of the capital city of Addis Ababa. There were two feeding centers established south of Addis. To support these feeding centers, two depot complexes were established: one in Kombolcha in the north and the other in Nazareth, to the southeast of Addis. The Nazareth complex was established not only for the convenience of serving the projects in the south, but it was also on the Djibouti-Addis railroad to accommodate the food commodities that were expected to be diverted from the Port of Assab to the Port of Djibouti and transported by rail to Nazareth and Addis Ababa.

Eventually, an experienced logistician was assigned as the logistics manager, reporting directly to the project director. The logistics manager is responsible to the project director for effectively performing the logistics functions and

One of two Twin Otter aircraft in Ethiopia used to haul supplies and personnel to and from World Vision's camps and feeding centers.

for providing direct logistics support to the project. A major lesson learned in Ethiopia was that an experienced logistics specialist must be assigned to the team early in the project.

2. **Administration and Flight Operations**: World Vision owned and operated two Twin Otter aircraft used to haul supplies and personnel to and from the camps. The logistics manager established an administrative section which monitored and coordinated the flight operations in support of the project. The administration chief was responsible for processing all the paperwork that came into the department. Other administrative responsibilities included maintaining files, preparing reports, keeping records of logistics personnel and performing other routine administrative duties. The flight operations group scheduled all flights. In December 1986, the two Twin Otter aircraft were sold when the emergency relief project began to phase down.
3. **Procurement and Port Clearances**: There was a shortage of qualified logistics personnel, particularly among the expatriates. Consequently, there were two rather unrelated functions--procurement and port clearances--which were combined and assigned to an expatriate qualified to manage both functions. Local nationals were assigned to assist and be trained in each of these functions. The procurement function involved both local and international procurement. Many needed items were not available on the local Ethiopian market. Consequently, international purchase orders were placed in England and elsewhere in Europe. Maximum utilization was made of the free ocean freight provided to the Private Voluntary Organizations (PVOs) by the Band-Aid group who had been successful in fund raising in Great Britain.

   As ships arrived at the Port of Assab and were to unload World Vision food commodities, it was necessary for qualified personnel from the logistics staff to meet these ships to ensure that the cargo was properly cleared, received and disposed of at dockside, either loading cargo directly into World Vision trucks or placing commodities in temporary protective storage.
4. **Supply Management**: The supply management function is to maintain the proper accountability of

all supplies and materials including the vast tonnages of food commodities. This particular function was handled by an all-Ethiopian staff. Nationals who had prior accounting experience performed extremely well in this function. A more thorough explanation will be given to this function in Chapter IX, "Materials Management." Initially, during the crisis period of the relief project, there was inadequate accountability of the food commodities. At the request of the World Vision Ethiopia staff, assistance in the area of stock control was provided from the International Office. An advisor, knowledgeable in setting up stock control systems, worked with the WVE staff and, together, they established and set into operation a complete centralized stock control system with the use of stock record cards. The system has become a model for other projects to follow.

5. **Commodities and Warehousing**: By necessity, World Vision plans to establish commodity manager positions in all projects where U.S. government-furnished commodities are being provided. This will be explained in Chapter XIII, "USAID-Sponsored Programs." There are specific regulations and accounting procedures set forth by the U.S. government in handling PL480 commodities. This requires an individual on the World Vision staff who is knowledgeable of these regulations and procedures. The reporting part of this job is normally not so demanding that it requires full time for one individual. Periodic reports are required to be completed and submitted through appropriate channels to the U.S. government offices in Washington. The basis for the accountability reports is mainly the records established and maintained in the formal stock control system. In the World Vision Ethiopia logistics organization, one expatriate was assigned a dual responsibility of managing food commodities, in addition to managing the two depot complexes at Kombolcha and Nazareth along with the other World Vision-operated warehouses located at the camps and the six warehouses which were located in Addis Ababa.
6. **Transportation**: From all of World Vision's relief projects in Africa, the transportation and distribution function has naturally been a key factor in logistics. (Chapter VI will be devoted to the entire spectrum of transportation.) The truck transportation in each country project has, by necessity, been handled differently. As general guidance to you in the field, the hard and fast policy of the International Office toward truck transport is that wherever possible the truck haul should be contracted out. The purchase of trucks is discouraged. World Vision does not want to be in the truck transportation business if it can possibly be avoided, because it causes high capital investment costs and it distracts from our primary field ministry of bringing relief to depressed people.

   It has not always been possible to avoid purchasing our own trucks, and it has caused many problems. In Ethiopia, an early attempt was made to rely on the government-owned trucks to distribute our food commodities. Unfortunately, these trucks seemed never to be available when they were needed. Consequently, World Vision was forced to purchase and establish its own truck fleet. Long-haul trucks with greater carrying capacity were used to haul from the Port of Assab to the depots. Short-haul trucks, capable of maneuvering in rough terrain and carrying less tonnage, were used between the depots and the camps. World Vision's capital investment in this fleet was quite large. Our purchase of trucks and vehicles in Ethiopia could not be avoided. An initial agreement was made with the government of Ethiopia that these World Vision-owned trucks and vehicles would revert to their control at the termination of the relief project. This transfer began in early 1987.
7. **Maintenance**: A plan was conceived for World Vision to contract the operations and maintenance of the Ethiopian fleet, but the plan was abandoned when approval could not be obtained for an outside contractor to move in and provide these services. With this decision there was no other alternative except to make arrangements to handle these transportation and maintenance functions ourselves. These responsibilities, of course, fell to the logistics department. By assuming these responsibilities, it meant that provisions had to be made for repair shops, mechanics, repair parts, tools and everything else needed to maintain this truck fleet, along with the large number of administrative vehicles. The control of the administrative vehicles and the implementation of a scheduled maintenance program for these vehicles required close supervision and direction on the part of the transportation section. (The maintenance function is covered in more detail in Chapter X.)

The Ethiopia logsitics organization was specifically designed to perform the logistics functions required to support the relief project. In planning your organization, you should identify the major logistics functions that must be performed and fit the best qualified personnel available to fill the key positions. Where qualified personnel are not available, immediately initiate requisitions for approval and recruitment. In the meantime, make certain that the best qualified personnel are assigned responsibility for your key logistics functions. Use the model organization described here for Ethiopia and alter it to fit your specific needs.

Reporting Logistics Activities

1. **Reports**: In every relief project, you can expect to be required to submit periodic reports to document the progress of your activities. The International Office

will place a requirement on the field director to keep him or her informed concerning the project's progress. The director in turn has an obligation to keep the donors informed who are contributing the necessary funds and resources. As the logistics specialist, you are obligated to keep your project manager or field director informed as to the progress of your logistics activities. It is important that your management receive a report from you periodically, at least once each month, on the significant happenings of your activities. As an example, for the particular reporting period the field director may want to know such things as:

a. Tonnage by type of commodity received at the port.
b. Tonnage received and on hand at the depot, distribution centers and camps.
c. Tonnages distributed to the villages or camps.
d. Quantity of contaminated or spoiled food on hand.
e. Number of trucks and vehicles in service. Numbers on deadline.
f. List of goods and services (over a specified amount) purchased.
g. Any significant maintenance problems.
h. Any truck or vehicle accidents.
i. Available storage space in warehouse.
j. Quantity of commodities by type in open storage.
k. List of personnel shortages.

There may be other pertinent information that the field director will want to have included in your periodic reports. You have an obligation to keep him informed of any significant event as it occurs. If the event is critical, it should not wait for your report. All good staff officers will keep their management informed. Another benefit of these reports is that they will force your team members to stay on top of their activities. Also, reports give the field director confidence that his logistics team is on top of the situation at all times.

IV. Coordination

". . . for the joy of the Lord is your strength."
Nehemiah 8:10

National, Provincial and Local Governments

Before any survey has been made of the region to determine the area of greatest need, the project officer should have already coordinated with officials in both the national and provincial governments of the host country. In most countries the provincial governments are autonomous and can act independently from the national government authorities. Consequently, it is advisable to coordinate and work closely with the provincial authorities. You can accomplish more and encounter less obstacles if the provincial government is on your side.

It is particularly important that the logistics specialist appreciate the benefits that can be gained through good government relations. Field office representatives should personally contact officials at the national, provincial and local levels to establish good relationships. It is also important and helpful to the project for management officials from the International Office on their visits to the projects to pay courtesy calls to the top government officials and the leaders of the local churches. Much good can come out of these visits. These representatives can conveniently lead the interference and provide the necessary framework for the future working relationships of the project team.

The provincial authorities know where the areas of greatest need are in their region. The project team should include the logistician when conducting a survey of these areas prior to planning the operation. Once the team has established what must be done and has determined World Vision's available capabilities and resources, the logistics specialist should begin to line up the tools he will need to get the job done. Before making any World Vision commitments on costly capital items, such as trucks or radio communications equipment or warehousing, first determine what the government will be able to provide in the way of support.

An example of a good provincial and local government relationship with our project team took place during 1985 when there were 1,000 MT of maize donated by the U.S. government to be distributed in the 1st Region of Mali. The seat of this provincial government was Nioro. The maize was part of the barter arrangement the U.S. government made with the government of Ghana to provide 10,000 MT to the 1st and 7th Regions of Mali. The purpose of the barter arrangement--taking advantage of Ghana's first surplus crop production after several years of drought--was to get food into the remote villages of Mali before the rains started. Heavy rains were anticipated which would make many roads inaccessible for trucks. Because of unnecessary red tape, the start-up of the truck haul from Ghana was delayed over a month. In the meantime, the rains had begun in the 1st Region, leaving many hungry people who could not be reached.

Our project manager had established excellent working relations with the commandant of the region and his local officials in Nioro. After joint preparation of a detailed distribution plan for the region, the commandant was able to identify those villages that could not be reached. World Vision had only two of their own Magirus trucks available in Nioro, and one of these trucks was deadlined because of a lack of repair parts.

After the only road from the capital city of Bamako to Nioro was washed out because of the heavy rains, the maize which had begun arriving by truck from Ghana was loaded onto rail cars and transported to Kayes, a railhead in the 1st Region located west of Nioro. There was warehousing available in Kayes, and the one main road to Nioro was still open. With the cooperation of the com-

mandant and close coordination with the World Vision team leader, the provincial government was able to assist in the transportation by offering us several trucks.

This close coordination resulted in an ingenious plan to reach certain villagers who could not normally be reached because of the heavy rains. Using the commandant's radio communications, government officials notified isolated villages to go to designated distribution points where truck loads of maize could be delivered. The villagers responded by coming from all directions on foot, by ox carts, donkeys and any other means available to receive lifesaving food. Many lives were saved through the closely coordinated planning actions between the World Vision project team and the provincial and local government officials.

It is essential that the project have the full support of the local government at all levels. The government will control many of the resources needed by the project team to accomplish what must be done. The use of some of these resources, such as truck transport, will probably need to be contracted for and must be funded. Be certain that funds are available for these contract services. In the matter of contract format, Chapter XIV, "Contracts" will be devoted to discussions and procedures for contracting.

USAID

Where U.S. government-furnished food commodities are used or planned to be used in a project, coordination with the in-country USAID office is imperative. Coordination with U.S. Agency for International Development (USAID) at the Washington level is the responsibility of World Vision Relief Organization (WVRO). Initial coordination with USAID is made in the country being supported, and it is made very early when the project is being conceived and the authorization has been granted to make the initial survey of needs.

The main point of contact in the USAID Mission is with the Food for Peace officer. One person needs to have regular contact with the Food for Peace officer and have the authority to make immediate decisions affecting the project. This person should either be the project manager or the project coordinator who should be located in the capital city, in order that this coordination can be continuing and effective. USAID policies and regulations do not require that this coordinator be an American citizen; however, from the standpoint of convenience, it is desirable.

The Food for Peace officer and the World Vision coordinator work closely together in establishing requirements for the project. It is the responsibility of the coordinator to prepare the operational plan for the project. Much of the guidance and direction on format for this plan is provided by the Food for Peace officer. When he or she is satisfied with the plan and is comfortable with the quantities of food commodities requested, USAID approves the plan and forwards it to Washington for approval. A copy of the plan is given to WVRO.

As the operational plan is approved and eventually implemented, this close working relationship and coordination between USAID and World Vision become even more important. USAID can also be of much assistance to our field team in obtaining resources that are available from the host government. These could be resources essential to the project such as transport, facilities, warehousing and personnel.

Other Private Volunteer Organizations (PVOs)

In severely depressed areas, there are often several PVOs which assemble teams and resources to help bring some relief to the stricken villagers. Normally at the national level, representatives of these PVOs meet once each week to exchange information and to effect necessary coordination on matters pertaining to their relief missions. Many of these organizations have different ways of providing relief. Some concentrate on medical and health care, some in community development, some in food distribution or other areas. However, they all have one common purpose: they are first responding to the needs of the depressed people to save their lives, and then, finding out how they can help these people help themselves. This has been World Vision's prime objective in meeting the emergency food requirements in the severe drought regions of Africa. We may first give grain, but then we want to teach the needy how to sow and cultivate their fields.

Because of the common purpose of the participating PVOs, these groups benefit by cooperating and coordinating with each other. For those involved in food distribution and in the operations of nutritional feeding centers, the potential exists for organizations operating in the same general area to share warehousing, truck transport, maintenance shops or other typical logistics resources. The logistics specialist should become well acquainted with his counterparts in these organizations. Through close coordination and cooperation, the specialist can generally economize by sharing these facilities, thus avoiding costly duplication.

Again using the 7th Region of eastern Mali as an example: When World Vision first moved into this area in early 1985, there was only one other PVO with any significant capability operating there. This PVO was the International League of Red Cross. Our project staff realized the difficulties involved in providing support to this entire region which is over 400,000 square kilometers, larger than the state of Colorado. Hunger, famine and malnourished children were prevalent throughout the 7th Region largely among the Nomad tribes who had lost all of their livestock as a result of the drought; consequently, they had nothing to trade or sell for other kinds of food, which had generally been their custom. There were over 250,000 people, mostly Nomad tribes, who were in severe stages of hunger.

We made an initial effort to coordinate with the Red Cross, but the Red Cross did not seem to be too receptive. Red Cross's initial reluctance to cooperate was largely because we were slow in getting teams organized and in providing the necessary resources to do the job. The Red Cross lacked the confidence in our ability to be an effective partner in sharing the relief effort in the 7th Region. We recognized the importance of coordination; a truly cooperative spirit on the part of organizations could economize on logistics resources and provide much greater support in the region.

In the slow process of World Vision finally "getting its act together," certain and partial agreements were made with the Red Cross in an attempt to split the region. It was generally agreed that World Vision would concentrate its efforts in the Gao-Menaka areas in the south, and the Red Cross would provide support to villages in the north of the 7th Region. The relationship between the two organizations was good.

The importance of close coordination with other PVOs operating in the same area cannot be overemphasized. An agreement about the exact area of responsibility will tend to limit the scope of operations for any one organization. This will allow that organization to more effectively define what is needed in order to do a specific job. Close coordination and cooperation will greatly help to economize on the equipment and facilities and will prevent unnecessary and costly duplication.

A local labor team unloads food from a World Vision truck at one of our feeding centers in Ethiopia.

World Vision Relief Organization (WVRO)

It was mentioned that WVRO coordinates with USAID at the Washington level. Many of our contract employees at the field level do not understand what WVRO is and what it is organized to do.

WVRO is separately incorporated from WVUS as a nonsectarian, humanitarian agency. The U.S. government is not authorized by federal regulations to conduct business with a sectarian international organization such as WVI. To be eligible for food and cash grants from the U.S. government, the recipient organizations must be nonsectarian with a U.S. registration, such as WVRO. WVI's field ministries and projects are the ultimate recipients of these food and cash grants.

WVRO is registered with USAID and receives U.S. government food and cash grants on behalf of the World Vision Partnership. In addition to receiving governmental aid, WVRO is responsible for handling all other Gifts-in-Kind (GIK) contributions to WVUS. WVRO also acts as a liaison between WVUS and WVI, as well as represents World Vision to U.S.-based private and voluntary organizations. WVRO is governed by a separate board of directors and prepares a separate financial statement. However, WVRO operates in close affiliation with WVUS in the support of WVI ministries. In FY84 WVRO's total income was in excess of $25 million. Projected WVRO income for FY85 surpassed $72 million. This is the value of the food and cash grants made available for WVI field ministries.

Gifts-in-kind contributions handled by WVRO include all noncash contributions to WVUS. GIK contributions in 1985 included surplus grains from the U.S. government, medical supplies, vegetable seed, portable sawmills and other woodworking equipment, and some privately supplied foods. In the case of surplus foods donated by the U.S. government, our field project teams are responsible to ensure that these foods reach those people who are truly in need, with a minimum of loss, and that foods are not misused in any way. These GIKs require substantial monitoring and reporting requirements. In addition, GIKs are only accepted by World Vision when they have clear indications from our field project teams that these commodities are needed to support our ministry in that particular country.

WVRO is frequently offered private GIK. It is WVRO's policy to offer all such donations to WVI and, if there is no interest by WVI, to offer them to other PVOs. This is another benefit of our field teams maintaining close coordination and cooperation with other PVOs operating in the same country. This can invariably result in a mutual exchange or sharing of logistics resources and facilities.

WVRO is often represented in the field and is included as part of our field project teams.

It was mentioned previously in this section that handling U.S. government GIKs involves substantial monitoring and administrative reporting. These procedures are all outlined in lengthy and rather complicated regulations. Specialized training and instruction are required to qualify individuals in handling these administrative procedures. In the preparation of the Mali projects, WVRO arranged for an outside contractor, Marine Overseas Services, to be responsible for providing direct support to WVI field teams by handling all of the administrative reporting and providing experienced field personnel to assist the WVI field team as required. This assistance was to include setting up accountability procedures which would be compatible with the reporting requirements.

The contractor was also given the responsibility, funded by the U.S. government, to arrange for and monitor the truck transportation of the donated maize from warehouse locations in Ghana to the designated distribution centers in Mali. To assist WVI in accepting our own responsibility in future projects, and with the phasedown of the contractor's involvement in Mali, contractor representatives were designated to train WVI personnel on-site in Sudan and Mali. Ultimately, these trained personnel can serve as cadre to train others and function as commodity managers on future WVI relief projects.

World Vision Ethiopia was an example of ideal coordination and cooperation between WVRO and the WVI project team. WVRO assigned one of their staff who was experienced in handling U.S. government GIKs to the WVI project team. The program director in turn assigned this individual directly to the logistics department. With the experience and Christian attitude of service, this individual became an invaluable staff member to the logistics team. As the commodity manager, his primary responsibility was to handle the administrative reporting of the U.S. government GIKs. Since this was not a full-time job, he was assigned additional responsibility for supervising the two logistics depots, along with the operations of all WVI warehouses located at the 11 campsites and the supporting warehouses in Addis Ababa. This arrangement was extremely effective and greatly contributed to the eventual success of the logistics operations in Ethiopia.

Maximum coordination and cooperation by our project field teams with other agencies, appropriate government officers and others involved in the relief effort are essential to a successful project. It is not possible to operate in a vacuum and still be able to conduct a successful operation. This is particularly true for the logistics team. Since our funds and resources are quite limited, it is necessary to coordinate with other agencies to share resources and exchange information essential to the project. Much can be gained and saved through this coordination.

V. Procurement

"Rejoice in the Lord your God . . . He sends you abundant showers, both autumn and spring rains, as before. The threshing floor will be filled with grain."

Joel 2:23-24

For the purpose of this particular chapter, it is again important to emphasize that this handbook is not restricted to only Africa. The contents of this handbook are applicable to any geographical region where World Vision is planning and conducting major relief and development projects.

During the African drought crisis, there was no consistency in the organization for procurement (or purchasing as it is sometimes called). On certain teams, procurement was handled by the Accounting department. On others, the Personnel staff was doing most of the buying. On the Ethiopia team, the responsibility was assigned to the logistics manager, and he had a separate section that was responsible for both local and international purchasing. Some field teams asked certain support offices to purchase and ship specific project items (which did not work out satisfactorily and therefore is not recommended). Support offices can be requested to do the shopping to establish availability and costs of items, but the actual purchase requisitions should be initiated by the field team, pending further directives from the International Office.

In spite of the different methods and organization for handling procurement, and in the absence of firm guidelines for the field to follow, our field teams have done a good job and have been able to get the most for their procurement dollar. As the size and magnitude of projects increase, the amount of the procurement dollar becomes greater, thus increasing chances for mistakes, potential waste and improper purchases.

As for organization for procurement, it has not been determined where the procurement function is assigned in the International Office. But the procurement function should definitely be in WVI, so the experts can be easily accessible to the regional vice presidents, relief and development project coordinators and others who are responsible for field projects. In the field, it is advisable to assign this function to the logistics team. Where a logistics team has not yet been established, the procurement function should report directly to the project manager or field director.

In anticipation that more specific policies and procedures are needed for the field teams, a detailed questionnaire was sent out by the International Office over the signatures of the regional vice presidents. The purpose of the questionnaire was to find out exactly how these various field offices are organized for procurement activities and how they are handling both local and international procurement matters. At the time of publication of this handbook, some of the responses to the questionnaires have just been received and are being evaluated. This study will result in an improved reorganization for field procurement and the dissemination of more specific policies and procedures for future field procurement activities.

Some of the types of questions asked in the questionnaire are as follows:

1. What methods or resources do you have available to you in determining or locating who you will purchase supplies from?
2. Do you have any set of guidelines and procedures which you follow when purchasing?
3. What methods are used in determining the selection of supplies?
4. Do you have a trained and experienced purchasing officer on your staff?
5. Do you obtain prior approvals from any higher head quarters before placing an order? If so, what dollar limits, if any, are used?
6. What process do you follow in placing a purchase order?
7. What is the usual method of payment on a purchase order?
8. What methods of transportation are normally used in delivering supplies from an international supplier?

World Vision distributes food near a distribution center in Ethiopia.

In the absence of any specific guidelines from the regional or International Office, the field team can conveniently use the questions asked in this questionnaire to evaluate their current policies and procedures for purchasing. The objective of sound procurement practices is to provide maximum protection of our funds used in the purchase of supplies and equipment. Until more specific guidelines are disseminated by the International Office, the field offices should continue to use sound judgment and good common sense in handling the expenditure of funds for the purchase of project supplies and equipment. Procurement is a function that deserves and demands constant supervision and monitoring by field management. Always be conscious of the source of funds that make our projects possible. This is good stewardship.

Management must seek ways to best maintain control of the procurement activities. Buyers who have complete freedom can become carried away and can jeopardize a project through careless and reckless overspending. One simple way that management at any level can maintain control is by providing authorization and granting approval for expenditures over a certain specific dollar amount. This is called centralized procurement. The project managers or field director can enforce this restriction, or it is even feasible for the International Office to insist that they approve all purchase orders over a specified amount.

It is even possible for the International Office, or the regional field office, to establish a central purchasing section whereby all field purchase orders expected to exceed a certain dollar limit would be passed to this office for procurement action. If such a system of a central buying office is established at any level, it is even more important that whoever requisitions the item must include a complete set of specifications to the purchase order so that the buyer will know exactly what to look for. Another benefit of central buying is being able to take advantage of the lowest transportation and shipping costs.

Procurement activities are related to the support of the project, and most purchases are made in support of the other logistics functions; namely, maintenance, transportation, materials management, etc. In other words, the purchase of supplies and equipment, either locally or on the international market, are usually in direct support of the project. And the direct support of a project is a logistics responsibility. In order to ensure continuity and a standard procedure, it is advisable that procurement activities not be fragmented. Whether you are buying pencils, copy machines or vehicles, it is best to place the responsibility for all purchasing under one logistics section or procurement head.

Local Procurement

A logistics planner must identify the sources from which materials and supplies can be procured. As much as possible, supplies should be obtained in the host country, providing the costs are reasonable and the items meet project specifications. This is local procurement. Overall, supplies purchased from local sources should be less expensive than international sources because they do not have high premium transportation charges. From recent experiences in the Third World African countries, supplies are extremely limited on the local market, making it necessary to purchase out of the country. In some cases where certain supplies are available locally, they may be acquired either by donation or purchase, and the sources may be either governmental or nongovernmental. The individual in the logistics team who has been assigned procurement responsibility should, early in the project, survey the local sources to see what is available for project use.

Have a local national who is familiar with the local markets be a part of the procurement section. These locals know the customs, habits and language of the host country and can usually negotiate a much lower price than an expatriate. If a local store knows an expatriate is making the purchase, the price automatically goes up-- and there is usually no room for negotiation.

In this chapter on procurement, understand that the procurement referred to is not only for materials and supplies, but for services as well. On many projects it has been necessary to obtain the services of local truckers, mechanics, laborers, etc. You are to treat the procurement of services exactly the same as you would the procurement of materials and supplies.

When you have the time and when there are multiple suppliers on the local market, it is always best to go out for competitive quotes. This is a procurement phrase that simply means to get two or more suppliers to bid prices for the supplies or services you need for the project. Competitive bids always get the best results for the most realistic and reasonable prices. Of all of our African relief projects, the most frequent mistake made by our field teams has been that they have been in too big of a hurry to buy; they take the first price offered by the first and only supplier solicited. This is in violation of the Field Financial Manual. Invariably, they then have ended up paying far more than they could have paid if they had taken only a little more time and had given more suppliers an opportunity to bid. When a supplier knows that he is the only one who has been contacted, his initial price is usually way out of range. If the buyer has not checked with other suppliers, how is he to know what the standard going rate or price is in the local market?

Competitive quotes do not need to be in a formal written bid offer. You can receive goods by a mere telephone call to other suppliers to compare prices. By so doing, it is possible to get a close approximation of the going rate or price. The few extra minutes that this action takes may save you from paying an abnormally high price.

When only one supplier is used, this is called a sole source contract. Sometimes it may be justified to go sole source, particularly when the supplier has the best quality that meets our specifications. In this case, you must know what the going rate is on the local market and be prepared to negotiate. Be cautious about going sole source, using this means only when the situation is totally justified and you have all of your bases covered. In most cases, there is no substitute for competitive quotes. If you are certain of the going rate or costs, it may be possible to get one particular supplier who provides quality items who is willing to guarantee his price for a given period of time.

In competitive bids the purchase order, or contract if necessary, should in principle be awarded to the lowest bidder. But it does not always work out this way. There are other factors that must be considered, namely, the quality of the product or services and the timeliness of delivery.

In Zimbabwe when preparing to receive 10,000 MT of maize from the United States through South Africa destined for Mozambique, we did not award the contract to the lowest bidder. There was only one bidder out of the five solicited who had the capability and willingness to do what had to be done (i.e., who would provide warehousing and truck haul and would haul the food commodities into hazardous areas in Mozambique). The quality of this particular company's operations was high, and their "can do" attitude was impressive. All of these factors were weighed before awarding the contract--even though they were not the lowest bidder. After World Vision made the selection to go with this particular contractor, subsequent meetings resulted in serious and successful negotiations since some of the incremental rates were able to be reduced to our benefit.

When purchasing goods, obtain the services of an independent surveyor to verify the quality of goods. On several relief projects, we have purchased seed to donate to families in impoverished areas. It is important that seed meet certain minimum standards of germination, purity, etc. An independent surveyor will, for a price, make necessary analysis of the seed. This type of outside service is to our benefit because of the protection it offers. Keep in mind that your team members are not experts in goods and services, and there are certain merchants who will take advantage of buyers who are not knowledgeable in their business. The independent surveyor, or an inspector who is an expert, will protect you and represent your interests.

Timely payment to the vendor after he has successfully supplied his goods and services is the best way for us to maintain our reputation and show that we are a Christian organization. Paying on time is a very sensitive issue and must be given proper attention. So many organizations or agencies become engrossed in complicated finance and accounting procedures, and the vendor who has given his best to please is the one who must wait for his payment. The agency, in turn, gains a bad reputation; the word will soon get out to other vendors in the community, who will then refuse any future business with that agency. The local purchase of goods and services is relatively straightforward, provided necessary funds are available and their expenditure is authorized.

As a rule, down payment for goods and services is not recommended. However, a vendor may insist on a payment of a certain percentage of the total cost when the goods are ordered, and the remainder of the payment on delivery. Payment is subject to verification of quantity and quality of the goods. When the vendor has met his obligation and the goods or services purchased are satisfactory, we should be timely in our payments. The buyer who has purchased the goods or services cannot "pass the buck" to the finance and accounting department if payment is not timely. It becomes the moral responsibility of the logistics representative to follow up and ensure that timely payments are made. In situations where the vendor has not fulfilled his obligations, payments should be withheld until the problem has been resolved.

In preliminary agreements on payment, it is very important that a specification is made whether the payment will be made in local currency or U.S. dollars.

Sudanese laborers load trucks from a warehouse in Wad Medani.

Most Third World countries benefit tremendously from payment in U.S. dollars because of the favorable exchange rate. Whatever currency is used, make absolutely certain it is agreed to before the order is filled. It is best to put this agreement in writing because of its sensitive nature.

World Vision learned an important lesson when purchasing trucks for a major project. Our staff was of the opinion that payment would be made in the foreign currency, but the supplier expected payment in U.S. dollars. There was no written agreement about what the currency would be. In the meantime, the World Vision employee who negotiated the sale had left the company, and we were "stuck" in making up the difference caused by the fluctuation in the exchange rates.

The purchase agreement or contract should specify the exchange rate to be used. Where foreign currencies are being used, it is possible and recommended that the finance office buy sufficient currency for the project when the exchange rate is in our favor.

Also, when ordering goods or supplies, specify any special packaging or labeling. These instructions are normally included in the specifications, but they should be clearly understood by both parties.

Essential to any type of procurement is the verification that adequate funds are available in the budget before the purchase is made. This could be very embarrassing to the procurement office if a purchase order is initiated and the order is filled, only to find out later that there were no funds allocated for the purchase. The procurement officer and the accounting department need to coordinate closely to ensure that this does not happen. Normally, the project manager will place a dollar limit on the purchase authorization so that no purchase order over a specified dollar limit will be initiated without the approval of the project manager.

When a buyer places a purchase order for any type of supplies or materials, the specifications must be made clear to the vendor. This is particularly true for repair parts or any other mechanical components. (More detailed information will be given on this in the following "International Procurement" section.) For local procurement, the specification is made simple because the buyer and user will be able to physically inspect the item before it is purchased.

Whenever possible, project materials and equipment should be purchased locally, if for no other reason than to save transportation costs. Our Ethiopia team encountered problems when they purchased typewriters, copy machines, calculators, etc., from sources in Europe. It was learned later that repair capability and a source of spare parts were not available locally. It would have been very costly to return these items to Europe for repair. It was possible that some of these items could have been purchased locally.

Particularly, when buying mechanical items locally, make certain that spare parts and components to the major item are available, and repair facilities can keep the items in good working order.

International Procurement

Most of the procurement for the relief projects conducted in the Third World countries of Africa has been, by necessity, done outside the host country. This is commonly referred to as international procurement. Locally, the teams have been able to purchase office supplies, oils and greases, food, building materials, utensils and many other common items used in the day-to-day operations. For items not so common, it is understandable why such items would not be available in these countries.

Early in the project, the buyers must identify sources for purchasing the items needed. Key suppliers and purchasing agents must be contacted who can respond to a purchase order. For Africa projects, European sources have been used the most because of their availability, quick response, lower transportation costs, etc. When they are in agreement, World Vision support offices can assist the buyers by determining the availability and costs of items in their country. However, the field procurement section should initiate the actual purchase orders and follow through with the transaction.

It is important for the procurement section to have immediate access to the key suppliers. Initial contact should usually be made by phone or telex to ascertain the availability of needed items, estimated shipping time and costs.

Depending on the type of materials being ordered, appropriate specifications should include:

1. Brand name and model number, where applicable.
2. Parts catalog number (be certain to make positive identification of the part in the catalog).
3. Quantity and quality of the materials or supplies.
4. Detailed description of the items, such as size, weight, composition, etc.
5. Country or region where the items will be used.
6. Technical or operational features of equipment.
7. Packaging, packing and marking instructions including exact title and address of the consignee.
8. Transport and shipping instructions.
9. Terms and mode of payment.
10. Any cancellation provisions.

In turn, the procurement specialist should require prospective suppliers to provide a detailed specification of the item and, where possible, to submit a sample. Essential information from the supplier should include:

1. Origin of goods.
2. Quantity that can be supplied.
3. Unit and total prices.
4. Packing in which the goods will be delivered.
5. Conditions and expected date of delivery.

In 1980 children in Kampuchea who had lost their parents at the hands of the Khmer Rouge wait for a meal at an orphanage.

6. Proposed terms of payment.
7. Guarantees as appropriate.

The procurement specialist should be especially cautious about purchasing costly capital equipment (i.e., nonexpendable equipment such as vehicles, portable warehouses, generators, etc.) If there is no agent or dealer in the host country, and the buyer is not all that confident about the specifications, it is often best for a member of the project team who knows what is needed to visit the supplier and personally inspect the item. This may be the only way the team can be assured that the item will do the job. Travel funds would be required. However, to spend a comparatively insignificant amount on travel is far better than buying an item sight unseen only to discover after it is received that it does not meet the desired specifications. It is also advisable to get some type of guarantee from the supplier.

A similar situation to this occurred to the Mali project team when they needed two portable warehouses to store the maize coming up from Ghana. A brochure was available from a firm in Norway that was selling fabric warehouses which could be quickly erected once they arrived in Mali. The purchase order was placed by telex from the information in the brochure. When the two warehouses arrived, they were considerably smaller than what was ordered. They were erected for urgent use. Soon after they were placed in use, the extreme heat inside caused a natural breeding ground for insects which attacked the grain, which was difficult to get under control. Had a member of the team personally visited the manufacturer in Norway, more would have been learned about this particular type of warehouse that may have avoided a purchase. Time was critical to the team, since they urgently needed some protection for the grain. These warehouses did serve that purpose, but they did have other faults.

There is no substitute for personal contact with the supplier or manufacturer when purchasing expensive items of capital equipment. When negotiating the purchase of the item, be certain that you have a complete and detailed list of specifications. Make sure that the item meets these specifications before finalizing the purchase. Where possible, buyers of more expensive capital items of equipment should be permanent employees who are familiar with International Office organization and procedures.

An agreement with the supplier on the method of shipment of the item is a key factor when negotiating the purchase. For much of the international procurement done for Ethiopia from European sources, shipment was largely by commercial vessels chartered by Band-Aid, at no cost to World Vision. The buyer must research the most economical means of transport. The supplier's shipping department should be able to assist in this research and make recommendations. At some ports, one shipping line may be given preference in berthing and unloading. This is another reason to establish personal contact with the supplier, and it is worth the travel costs. Freight forwarders can also be useful in providing the most economical shipping.

When the project team ordered Magirus trucks from West Germany for projects in Mali, the team decided to drive these vehicles through Europe, ferry across the Mediterranean, and drive across the desert to the project location in Gao. This trip took its toll on the trucks. Several mechanical discrepancies had to be corrected before the trucks could be used on the project. Before making the trip, the team did not compare driving the trucks versus using ocean freight to a port, possibly Abidjan in Ivory Coast, then driving over improved roads to Mali. It was quite costly to pay the Norwegian drivers and cover their expenses during the trip from Germany and their return trip back to Norway. The trucks across the desert did save some time, which was an important consideration. But the trucks showed much wear and had many mechanical problems after they arrived which were not easily corrected.

This is all the more reason for the logistics team to do its planning in advance so that needs can be anticipated and actions taken that are the most expedient, practical and economical to the project. Proper scheduling is all-important. Do not save the procurement of equipment essential to your project to the last. Anticipate your needs and take necessary, timely action to fill them.

In screening international suppliers, most requests for quotations will be made by telex, especially in the early stages of an operation. These requests should include the following: separate figures for the cost of goods, export packing, insurance, freight, transport (air and sea), any special packaging required, customs documentation, estimated shipping time and other essential information that the buyers can evaluate and determine the best qualified suppliers.

When the supplier has been selected and the cost negotiations have been completed, one of the first orders of business is to agree on the method of payment. The supplier naturally wants to know how and when he will be paid. Some options of payment are:

1. Payment on receipt of an invoice after the project team has taken possession of the goods.
2. Transfer of funds at various stages agreed to by both partners.
3. Cash when the order is placed.
4. Payment by a letter of credit.

Whatever method of payment is agreed upon by both parties, it is our responsibility to ensure that payments are made according to this agreement and that payments are timely. Again, it is essential to agree in writing the type of currency that will be used in payment. When in doubt as to the best method of payment to use, consult the procurement specialists available in the International Office for assistance.

For international procurement, expeditious processing and clearances at the receiving port and uninterrupted transport through transiting countries are vital factors when purchasing goods and equipment. The success of shipping, receiving and processing goods is accomplished by proper documentation. Particularly on major items, this is part of the service that is provided by the supplier. Large international companies or manufacturers will have shipping experts who should be familiar with the various country regulations and will know what proper documentation is required. It is the responsibility of the relief agency to ensure that all necessary waivers are obtained from the host government to allow these goods entry, free from the payment of any customs duty. The agency should obtain these waivers in writing. Proper documentation is essential in avoiding unnecessary delays and tie-ups when filling international orders.

When items are procured from funds authorized by USAID grants or cooperative agreement, there may be additional restrictions involving the purchases. USAID grants normally require that use of funds be limited to:

1. Items of U.S. source and origin.
2. Purchases made within the host country.
3. Purchases from developing countries.
4. Purchases from Western countries.

Logistics and procurement specialists should be alert to these restrictions and seek the guidance of the commodity manager or USAID country representative before placing purchase orders.

Until more specific policies and procedures are released to the field, teams should follow the Field Financial Manual. Field procurement activities have not received the emphasis in the past. However, because of the scale and magnitude of relief and development projects and the high costs of goods and equipment necessary to operate the project, field procurement warrants special attention. A staff of experienced procurement specialists are beginning to take their place in the International Office to handle these matters.

What department these responsibilities will fall under in the International Office is not all that important. But it is important to specify who in the field project team will have procurement responsibilities. Since procurement is a common function of logistics, the responsibility should be assigned to the logistics manager. Members of the logistics team who are given the responsibility of handling procurement matters will be expected to comply with the policies and procedures set forth by their next higher headquarters and by the procurement section of the International Office.

It is also important that the project manager or field director give his close personal attention to field procurement matters. This can be done by being the final approving authority for all purchase orders initiated by his team.

VI. Transportation By Truck

"Let them give thanks to the Lord for His unfailing love and His wonderful deeds for men."

Psalm 107:8

During the period of confronting the recent drought in Africa, the one function of logistics that provided the greatest challenge to World Vision's field teams was transportation. The problem of organizing the transport of relief supplies to the villages in bringing food to the hungry people was difficult and in some cases quite costly. Because of the remoteness of many locations which were established as objectives for relief commodities, it made transportation the most challenging logistics function to cope with. As God answered prayers for rain in many of these areas that had not seen rain for several years, He didn't seem to be overly concerned about the problems of inaccessible roads and washed out bridges that created distribution problems for the logistician. In most cases, the problems created by the lifesaving rains were overcome by proper advanced planning.

In relief projects where there are a large number of tons of food commodities to be distributed, the project logistics team should give priority to the assignment of a transport coordinator. Regardless of whether or not we are operating our own fleet of trucks, as in the case of Ethiopia, or leasing trucks as in Sudan, a transportation specialist is invaluable to the team, having the primary responsibility of coordinating all transport operations. There are many other problems and related responsibilities when World Vision elects to purchase trucks and operate its own truck fleet. Among many of these related responsibilities are maintenance, driver training and control, provision of spare parts, fuel, workshops and insurance.

Trucks are not the only concern of the transportation section. In most relief projects we will purchase administrative vehicles such as land cruisers, pickup trucks and passenger buses. In most countries in Africa, it is not possible to rent o. lease these types of vehicles. The management control of administrative vehicles can be a difficult and time-consuming job.

For example, in Ethiopia World Vision owned 110 of these types of vehicles. Most of these vehicles were dispersed over a wide area, being located in more than 14 various camps, depots and feeding centers. Since no maintenance facilities were available at the camps, a centralized maintenance facility was established in Addis Ababa. As a part of a routinely scheduled maintenance program, these vehicles would be rotated into the shop in Addis Ababa. This required effective and efficient control procedures.

Later in the handbook, a separate section is devoted to a related logistics function of maintenance. Because of the shortage of logistics team personnel, this related function will invariably be the responsibility of the transport coordinator. In projects where there are a large number of vehicles and equipment involved, an experienced maintenance coordinator, if possible, should be assigned to the logistics team.

During the drought crisis in Africa, there was an unwritten but recognized management policy of the International Office: project teams were not authorized to purchase trucks if there were trucks available in the country for rent or lease that could do the job. The purchase of trucks is a very costly capital investment. Normally, a relief project will be in place for a specific duration. Consequently, the eventual disposition of the trucks when the job is completed becomes an issue. In most cases the host country will require that the trucks remain in-country after the project is completed resulting in a rather costly, short-time investment for World Vision. In addition, by owning our own trucks, we assume full responsibility for the control, maintenance and support of the trucks. Project problems are compounded when the first truck is purchased. It is much simpler and cheaper in the long run if the trucks can be

A long-haul truck arrives in Gao, eastern Mali, with a load of maize from Ghana.

rented or leased. The owner assumes full responsibility for the maintenance, spare parts, driver control, liability and, in most cases, fuel supply--depending on the provisions of the contract. When we do own our own fleet of trucks, which in some projects is the only acceptable alternative, all of these additional responsibilities are inherited by the project team.

It was necessary for the Ethiopia team to purchase their own fleet since it offered the only solution to both the long- and short-haul distribution of the vast tonnages of food commodities received at the Ethiopian and Eritrean ports, which helped to sustain the lives of thousands of hungry people. Until we bought our own, the only other trucks available were under control of the Ethiopian government, and these could not be counted on for humanitarian use. The first priority use of these trucks established by the government was in support of their military forces which were fighting a civil war.

In contrast, in Sudan there was an abundant supply of trucks available for rent or lease. There were numerous truck firms with the larger capacity trucks used in the long haul from Port Sudan to our two designated distribution centers. Through the use of competitive bids, it made rates for long-haul truck leases quite reasonable. For the short haul, each of the major cities and towns had an organized "union" of local market (SOUQ) trucks available. The rates were somewhat standard but fluctuated depending on the truck's job demand. The local market trucks' range in capacity was from 5 to 8 MT, which is ideal for the short-haul distribution from the warehouse to the villages. Prior to distribution, the logistics team would negotiate rates and terms with the "union" leader. Logistics would specify the number of trucks available on a daily basis. The rates were based on a specified cost per day for each truck plus a cost per kilometer irrespective of load or distance traveled. The truck driver was paid after he returned from his trip with a signed bill of lading acknowledging delivery of the cargo. The driver was totally responsible for his truck and cargo, and the price included the driver furnishing his own fuel.

The arrangement greatly simplified our role in the ministry of feeding hungry people. The logistics team merely worked through the union leader who issued all instructions to the drivers relayed from logistics. The drivers were extremely competent and knew how to maneuver the roads and desert track. There were no unnecessary delays since the driver's pay was based on turnaround time, which was his incentive for a quick return to pick up another load. The drivers were required to pay a certain percentage of each trip to the organized union as a management fee.

The theory behind the International Office policy against ownership of trucks is not only because of the investment cost; the additional responsibilities to support the ownership of trucks would greatly distract from our primary ministry of providing relief and comfort to the depressed people. These additional responsibilities demand more funds and more people and create many more problems.

A Magirus 20 MT-capacity truck with a 10 MT-capacity trailer hauls food in eastern Mali.

Long-Haul Transport

The transport of large tonnages of food commodities and supplies is separated into two categories--long haul and short haul. The larger the capacity of the truck, the more economical it is to move the cargo, especially over greater distances. In Ethiopia the large quantity of food moved, the condition of the roads and terrain between the port and the depots made it possible to maximize the carrying capacity. The DAF long-haul trucks purchased by World Vision had a gross carrying capacity of 35 MT. In addition, an equal number of trailers were purchased with a gross capacity of 12 MT. The trailers were pulled in tandem behind the trucks, which of course added to their carrying capacity. This provided an excellent capability, and considerable tonnages were hauled from the port in a reasonably short period of time. The principal challenge to the depots, which were receiving these trucks and trailers, was their ability to store the cargo. The available warehouse space at the depots was limited; consequently, it was necessary to keep the commodities moving to the camps. With the number of people they were feeding, it was no problem to consume or distribute the food once it was delivered to the camps.

In the delivery of the maize from Ghana to distribution centers in Mali, 35 MT capacity trucks was the maximum limit. There were difficult places along the various routes that were desert track, making these places inaccessible to larger loads and not allowing trailers to be pulled in tandem. Also, for the trucks coming from the west en route to Gao, it was necessary to cross the Niger River. This river is a major obstacle, slightly over a mile wide.

There had never been a bridge in this part of Mali across the Niger River. Initially, a small, dilapidated ferry--less than 20 MT in capacity--sometimes was available. This ferry was out of operation more than it was in. Before the major transport of maize from Ghana began, our team hauled small truckloads of food from Bamako, nearly 1,000 miles, only to find the ferry not running. In order to get food across the river to meet the crisis, small canoes and boats were rented from local fishermen, and the bags of grain were shuttled across the river. This was a very slow, tedious process, but the only solution available at the time. The bags of grain were unloaded on the far side of the river where leased trucks took the bags to warehouses in Gao. Such a solution for handling a few hundred tons was acceptable. But, later, when the total quantities programmed for the Gao area were to be about 5,000 MT (or some 110,000 bags), this process was much too slow to avert a famine crisis. A better solution for getting the commodities across the river had to be found.

The U.S. government was made aware of this problem at the river. They had definite interest and a sizeable investment in this project. It was the government's responsibility to see that the food was delivered to the designated warehouse loca-

tions on the east bank of the river in Gao. The American Embassy made arrangements in Bamako to obtain a detachment of U.S. Army engineer troops from the NATO forces in West Germany.

These troops, along with their equipment, were flown into Bamako and shuttled by C130 aircraft to Gao near the banks of the river. Their barging equipment included large outboard motors and rubber pontoon boats. On top of the boats, a flat wooden deck was assembled to carry the trucks. The assembled ferry, designed to carry up to 90 MT, was propelled by large outboard motors. One trip across the river hauling two, sometimes three, long-haul trucks was completed in about 10 minutes. This engineer boat detachment remained on the river for about eight months, and it proved to be an effective solution. The troops received excellent training, and they became a part of World Vision's mission of mercy of keeping starving people alive. This allowed time for the Mali government to build a larger ferry for later use. The troops were eventually released to return to their parent unit in West Germany. It was a great spiritual morale boost for these young American soldiers to take part in this relief project; they were highly motivated for this job.

Short-Haul Transport

The classification of short-haul trucks is any vehicle that can carry fewer bags up to about 8-10 MT. Normally, the short-haul transport is our responsibility since the deliveries of food commodities to the receiving villages originate from our storage warehouses at the distribution centers. World Vision normally assumes possession of the food commodities at the distribution centers.

It is emphasized once again that the best solution for short-haul transport is to contract local truckers. This is the preferred solution over buying our own fleet. But it is also important to evaluate the available contractor's fleet to determine if his trucks can do the job.

To get to many places in the relief of hungry people, deliveries have had to be made to the "bush" where no roads exist. Normally, operating across desert track four-wheel drive (4WD) trucks are essential. In most cases, commercial truckers do not operate 4WD trucks. There is not the normal demand, and 4WD trucks are more expensive. As a result, the conventional truck will not go to places where the food commodities are most needed. You must have the type of vehicles necessary to get you there and back.

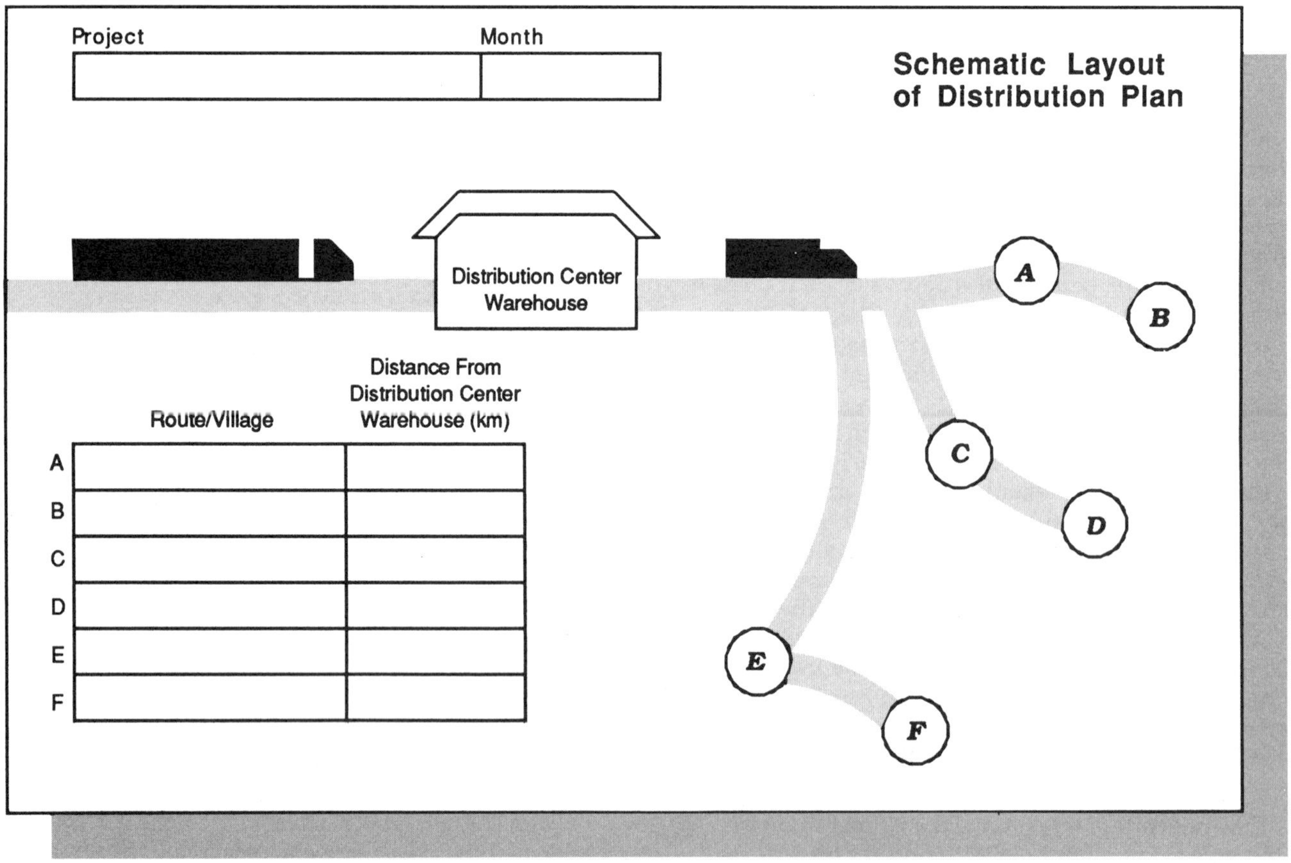

FIGURE VI.1

Some of these locations are so remote that, if a truck became stuck or stranded while en route, it might be days before the truck and driver could be found. For this reason it is always desirable to send trucks out in pairs; although this may not always be possible. The transportation coordinator should always be alert to the possibility of losing a truck and driver. Plans should be made in the event any truck is overdue on its scheduled return. Radio communication is the ideal solution under such circumstances, but this is an expensive luxury we can seldom afford. On one occasion in Mali, it became necessary to call on Mission Aviation Fellowship (MAF) to spot the location of a stranded truck by aircraft. If all else fails when sending trucks out to remote locations, at least send an assistant along with the driver with emergency rations and water.

Distribution

In preparing to make short-haul deliveries from the distribution center to the villages, the transportation coordinator should first work up a detailed transportation or distribution plan. Usually, the number of village locations designated to receive food are so great in number that a schematic layout is not practical. Figure VI.1 is an example of what a schematic layout might look like. It does help to visualize what must be done, but there is no way to include all the essential information of a distribution plan on a schematic layout. This information can best be shown in tabular form. A sample format is Figure VI.2. Ideally, a combination of both the schematic layout and tabular form should be effectively used. It is advisable, of course, to keep it simple and not to complicate what is attempting to be shown or displayed. A distribution plan is designed to tell you as much as possible what is required to be done and what all is needed to get it done. Essential information in a distribution plan would include:

1. Villages to be supported along each route.
2. Distance in kilometers from distribution center warehouse to each village being supported.
3. Number of bags required each month for each village.
4. Number of trips per month per truck to each village.
5. Amount of fuel required per month.
6. Number of trucks and drivers required each month.
7. Average load per truck per trip.
8. Estimated costs.

The villages to be supported are normally identified early when the initial survey of the area has been made.

Commodity Distribution Plan
From:

(Distribution Center)

Project

Month

To: Village	Route	Distance From Distribution Center Warehouse	Total Number of Containers/Month By Commodity Type	Number of Trips/ Month	Number of Trucks/ Month	Number of Containers/ Truck Load/Trip

FIGURE VI.2

In Sudan there were over 500 villages supported out of each of the two distribution centers, Wad Medani and Damazin. Each of these locations required resupply at least once each month. At one time some of the villages were resupplied more frequently, even two or three times a month. This placed a tremendous burden on the logistics organization. But the main concern of the project team was that if too much food were deposited at these villages at one time, the villagers would not be able to control the consumption. Because of the impending logistics problems created, it was necessary to advise the villagers that they would be resupplied once each month, and it was their responsibility to see to it that they had adequate food to sustain themselves until the next supply.

The convenience of the villages and the logistics requirements are not always compatible. In certain locations and at certain times, it is necessary to deliver up to four or five months' supply to the village at one time in order to sustain it during the rainy season when roads become inaccessible. The villagers must ration and manage their food to prevent running out before receiving another supply. It becomes a matter of discipline on the part of the villagers. A strong village chief will enforce this discipline.

When so many villages are being resupplied each month, it is important that the transportation coordinator make up a daily distribution plan. This is the only way accurate records will be maintained to show the frequency of distribution. A daily plan made up the day before will identify the number of trucks and drivers required and the number of bags to be delivered. Of course, the receipted bills of lading will confirm deliveries made each day. Bills of lading are explained in more detail in Chapter IX, "Materials Management." It is not always possible for a World Vision employee to actually monitor the deliveries as they are made to each village. However, in due course of time, a monitor should attempt to visit each village at least once each month to evaluate the food supply and condition of the villagers, as well as the food security and actual dispensing of food to the villagers. Much can be learned from these monitor visits. The monitorship of the food distribution is required when handling PL480 (U.S. government) commodities. Regardless of the source of the commodities, our company policy is to monitor distribution as closely as possible. It is too easy for the food to get into the wrong hands if the distribution is not monitored effectively. In countries such as Mozambique where there is heavy fighting because of rebel activities, it is not always possible to monitor the distribution in certain locations.

What singles out World Vision and its effectiveness in the field, is its policy of having an efficient, workable monitorship system to ensure that the food is received by the depressed people. Careless distribution can result in the food getting into the wrong hands. An effective system of monitoring is essential.

Administrative Vehicles

Every field project requires administrative vehicles. A project cannot be effectively run without these types of vehicles. They are generally not available to be rented or leased. However, in Sudan our team initially leased four Toyota Landcruisers on a month-by-month basis until arrangements could be made for our own purchase, shipment and necessary customs exemptions. When a team moves into a project, it immediately needs local transportation. In fact, when the initial survey group arrived in Khartoum in May 1985, the first action was to hire a few local taxi cabs with English-speaking drivers on a day-by-day basis. It is just not possible to get a project off the ground without local transportation for the staff. There is so much coordination that must be done. Because of the high cost of renting or leasing these types of vehicles, purchase arrangements for our own fleet should be given the highest priority. Determining the number of vehicles required is a management responsibility. Logistics should get into the act when ordering the vehicles and deciding on additional accessories and equipment to purchase for the vehicles.

The most popular and versatile administrative vehicle used on field projects has been the Toyota Landcruiser or the Nissan Patrol. The project team must first evaluate their needs and the purpose for which administrative vehicles are to serve. On certain projects pickup trucks have proven to be very useful, and in some situations the twin cab Hilux (with a pickup bed) provided transport for nutritional teams. These vehicles carry up to a ton of food commodities and equipment used to set up feeding stations in remote villages. The pickup is a handy all-purpose vehicle, particularly where there is a need to haul small quantities of food. This is far more economical than using a

An Ethiopian driver in a World Vision-owned truck delivers food commodities.

A short-haul truck distributes food to a village in Ethiopia.

10 MT capacity truck to haul a few bags of grain. For some extremely remote villages, the pickup is the only type vehicle that can get there and back.

Some field teams have purchased four-door sedans, largely as a status symbol for the team. They are not practical vehicles to have in the field, and their use is limited to city driving. The Landcruiser is a far more versatile and practical vehicle which can be used both for city driving and field trips. Often, monitor teams are required to take long field trips. The Landcruiser is a good vehicle for this purpose.

Field conditions and terrain will largely dictate the need for 4WD vehicles. In the 7th Region of Mali, for example, every vehicle wisely purchased had 4WD. Since there were no roads and all the traveling was done along desert track in sandy soil, the 4WD vehicle was essential. The commercial trucks available for lease in Mali were all standard two-wheel drive and literally restricted to the few areas where there were improved roads. Consequently, we were forced to buy 4WD trucks to allow us to distribute to the critically remote areas in the desert. During the rainy season many of these desert areas were completely inaccessible to even 4WD vehicles. The desert in some areas can be a grim place to try to operate during the rainy season. Trucks have been known to be stuck up to the windshields, making recovery extremely difficult.

On relief projects, especially in the more remote areas, the matter of transportation can be the greatest challenge for the logistics team. Transport is essential if the food commodities are expected to get where they are most needed. In drought-stricken Africa the people were usually so weak from hunger that it was not possible for them to come to the food--even though in many areas they were so desperate they tried, many did not make it alive! The transportation functions become so involved that on a large-scale relief project involving large tonnages for distribution, the logistics specialist should make every attempt to obtain an experienced transportation coordinator. Transportation does not merely include truck and vehicle transport. The next chapter will be devoted to addressing "Other Modes of Transport."

VII. Other Modes of Transport

"But thanks be to God! He gives us the victory through our Lord Jesus Christ."

I Corinthians 15:57

One of the first actions the logistics specialist must take in planning a relief operation is to thoroughly identify and evaluate the various modes of transportation that exist within the project country. If this country is a land-locked country (one that is located within the interior of the continent without having direct access to ocean shipping), it is particularly important to evaluate the available modes of transport to reach your project country. Normally, the freight forwarder who has been contracted to make all in-transit shipping arrangements to deliver your cargo to your distribution centers is expected to know the available modes of transport to your land-locked country. However, you are in a better position to evaluate these modes from the standpoint of costs, time delays, security and overall reliability. This is the only way you can be assured that the freight forwarder has done his homework and is providing World Vision the best and most economical service.

At one time during the drought crisis in the Sahel, WVI anticipated the possibility that we might become involved in a large-scale food distribution project in Chad. This did not come about because it was more feasible for us to obtain food direct from the World Food Program, requiring this agency to make the food available to the relief organizations from their warehouses in Ndjamena. Most of the seed that was distributed to hungry people in the south of Chad was purchased locally. In anticipation of direct involvement, we began to evaluate the various alternatives of transporting large quantities of food from a port to available warehousing in Chad. There was no simple and easy way. And this was found to be the case in evaluating all land-locked countries in drought-stricken Africa. Some ways were better than others, and it became a matter of weighing the alternatives, opting for those with the fewest and least significant disadvantages, with costs being a major factor.

One such alternative route that seemed reasonably practical, but which had disadvantages, was a road and rail combination along Nigerian routes using the Port of Lagos or Port Harcourt. A reasonably reliable railroad connects both of these ports with Marduguri, an average of about 1,600 kilometers distance from the ports, with Port Harcourt being the closest. From the rail terminus at Marduguri, the cargo would require transferring from rail to truck to complete the trip to Chad. The average transit time to Ndjamena (Chad) was two to three weeks. This routing was quite reliable, having year-round capability. The disadvantages included a high risk of loss en route, long port delays and difficult customs procedures.

The second most practical alternative, and the one most frequently used, was another rail and road combination using the Port of Douala in Cameroon. The average transit time was two to three weeks when the following factors existed: year-round capability, relatively simple administrative and customs procedures, comparative safety and reliability.

A third alternative (which was evaluated but not considered) was a transequatorial route involving a rail-water-road combination to Ndjamena. From the port of Pointe Noire in the Congo by rail to Brazzaville, then by the Obangui River to Bangui, in the Central African Republic, and then by road (truck) to destination in Chad. This rail-river-road combination required considerable handling of the cargo, resulting in an average transit time of four to six months. The risk of loss was high by waste and thefts, and the road conditions were poor. The long transit time made this alternative completely unacceptable. With such a delay in getting food to the stricken people, there would be no survivors!

In addition to the three alternatives mentioned above, there were three other possibilities that were not considered for various reasons. One was again through Nigeria, along the Benone River, which is only navigable two to three months of the year. Another was from Port Sudan by road and rail (which are not very reliable), and finally, from Libya, by mountainous road through a very unstable and troubled political area.

The main reason for devoting this time to the evaluation of the different modes of transport from an ocean port to the land-locked country of Chad is to emphasize the flexibility available in planning in-transit moves of food commodities. The major disadvantage of each of these alternatives was the need to transfer the cargo from one mode of transport to another. When you are handling 8,000 to 10,000 MT of commodities, this becomes rather costly and time consuming. Ideally, the transportation planning should, where possible, include an uninterrupted movement from point of origin (dockside) to final destination. There may be factors of economics, politics, customs, etc., that would prevent such an uninterrupted move, but all factors must be weighed before making a decision. Some freight forwarders will not weigh all these factors. This is the time when our logistics specialist must become involved and, if necessary, influence the decision.

Rail

In Africa, truck transportation has essentially been the predominant means of moving food commodities. In some areas, a railroad has been operating, but it is generally too inefficient and unreliable for it to be used. However, transport by rail is usually the most economical way of moving large quantities of relief supplies overland. Rail is also characterized by the capability to move large tonnages at speeds comparatively higher than inland waterways or truck transport. Of all the transport modes, rail transport is least affected by adverse weather. Rail transport, however, may be less flexible than other modes because it must depend on a fixed roadbed. Virtually, any commodity can be moved by rail subject only to clearance restrictions along the route. In countries where rail does operate efficiently, the logistics specialist should contact the railway authorities who sometimes transport relief goods free of charge or at reduced rates.

From experiences in developing countries in Africa where railroads do exist, they have generally been found to be too ineffective and lacking in reliable equipment. The management was poor and inexperienced to cope with the large build-up of tonnages to support the relief effort.

For example, it appeared that the best way to get relief supplies to Mali was to bring them into the Port of Dakar (Senegal) and rail them to Bamako, since there were no roads connecting these two places. After a complete investigation of the Senegal Railroad, it was learned that rail was too unreliable and terribly slow. The Senegal Railroad had been known to interrupt a shipment by unloading the cargo en route to its destination along a siding in open storage in order to transport cargo which usually has a higher political priority. (This is no way to run a railroad and in handling emergency relief goods, it is not worth the risk.)

At one time, there was a great deal of enthusiasm about diverting food commodities from Ethiopia to the Port of Djibouti and transporting by rail into Addis Ababa or along the route. This particular railroad is jointly owned and operated by both governments. Its outrageously high rates seemed to indicate its interests were more economical than humanitarian (to save lives). Also, its equipment was so outmoded and its operational know-how so lacking that the railroad could only accommodate about 300-400 MT per day at the most. The railroad would shut down without prior notification to the users.

If a railroad is operated efficiently, and it goes where you want to go, it is the preferred mode of transport for large tonnages.

Air

Air transport's main advantage is its ability to deliver quickly and close to the scene of the disaster. Its disadvantage is the high cost compared to the quantities

This Ethiopian woman receives a donkey, seeds and garden tools prior to returning to her village to grow her own food.

Foreign military forces in support of World Vision and other relief agencies in Ethiopia conduct an air drop of blankets from a C-130 aircraft.

(tonnages) to be delivered. In comparison, a C-130 Hercules aircraft can transport about 17 MT. The payload depends on the distance the plane must fly and the amount of fuel it is required to carry. It is also subject to technical limitations such as weather conditions, approach and landing facilities, ground and fueling services, load capacity, navigational aids, etc. Relief supplies can be sent by scheduled airlines (if available) or by chartered aircraft (which is generally the case). Often the local government will make their military aircraft available for emergency relief goods. The logistics staff must ascertain if this is possible before entering into any costly charter contract. In relief operations, it is very difficult for the logistics planner to weigh the most economical use of funds when famine victims are starving for lack of food. It is a great temptation to ignore the costs and charter an aircraft to bring relief to starving people.

World Vision experienced a similar situation in the early days of the relief effort in Tete Province of Mozambique. The famine in this region was at a critical stage. Maize was available in neighboring Malawi. Our representative planning the project had first-hand knowledge of the crisis because he had seen it. But by reacting too quickly, without taking the time to investigate other alternatives, and being influenced by the emergency nature of the situation, he hastily chartered an aircraft, and the air delivery operations began. Over 1,000 MT of maize was delivered to Tete City at a tremendous expense to World Vision. What made matters worse was that a railroad service was available to deliver the food to the same general location in Tete at considerably less cost! This is where the logistics specialist in charge of planning the support of a relief project must be alert and on his toes to avoid such waste of project funds.

In large-scale relief operations such as Ethiopia, World Vision found it necessary to invest in the purchase of two Twin-Otter aircraft. Because of the nature of the rough terrain, the unimproved roads and need for quick reaction, owning our own aircraft seemed to be the only solution. The aircraft contributed greatly to reducing the response time to emergency situations. These aircraft were configured to carry people, cargo, or both. By removing all seats, each aircraft could carry nearly one metric ton of supplies. These aircraft were used in their versatile role by transporting team members and other personnel from Addis to and from the camps and feeding stations. They were able to respond to emergency requisitions from the camps by transporting medical supplies and other essential items in support of the relief operations. Since these particular aircraft did not need elaborate, improved landing strips, they were able to get into isolated areas with food and medicine in situations where the villagers were not able to make it out to one of our camps. During the peak of the crisis in Ethiopia, these aircraft were "machines of mercy," responsible for saving many lives. Having your own aircraft is an expensive capital investment and costly to operate. These two aircraft were sold in December 1986 after the emergency relief project phased down.

Another use of air transportation in support of relief operations in remote areas is air-dropping food commodities. Air-drop operations were conducted in Ethiopia and Sudan by C130 aircraft flown by trained and experienced military crews. These operations were conducted in isolated areas where larger ton-nages of food were required and when permission was granted by the respective governments. Because of the rebel activity in Ethiopia, operations like this were not without an element of risk. In Ethiopia, reinforced bags of grain were free dropped (without parachute) on expendable pallets, at altitudes of less than 100 feet. After-action reports indicated that only a few bags broke open on impact. The operations were reasonably successful.

In the planning of such operations, the logistics team must provide close coordination with the military forces conducting the drop. It is necessary for the logistics team to provide a ground control party which can be flown in by helicopter equipped with radio communications. The main purpose of the ground party is to control the nationals around the drop zone and handle the distribution of the food after it gets on the ground. Such operations must be closely controlled and coordinated.

Inland Waterways

When the disaster area is served by inland waterways, the logistics team should make maximum use of this capability because it is an excellent means of economically transporting large tonnages of relief supplies. A good example is the country of Gambia in West Africa. The Gambia River serves as a major transportation artery and is navigable for almost the entire length of the country. Gambia does not have a fully integrated transportation network. There is no railroad and only one commercial airport. The main roads parallel the river. The majority of the road network is unimproved. The people of Gambia rely on the river as their primary transportation network.

In summary, truck transport is not the only means of delivering food commodities and other relief supplies. As was mentioned previously in another section, the logistics team in Sudan was ingenious in organizing a camel caravan with bags of grain loaded on the camels' backs to supply urgently needed food to an area that could not be reached at the time by any other means. A good logistician is one who can think and devise ingenious means to support a relief operation when there is no textbook solution to the problem.

In planning an operation, the logistician must evaluate all available means of transportation and make maximum use of what is available. Relief operations are temporary, and for that reason World Vision should not plan to make costly capital investments in equipment, which possibly cannot be utilized elsewhere at the completion of the operation. Normally, political pressure will prevent evacuation of the equipment, which means that the equipment will be turned over to the host country. In contrast, a development project could be of a longer duration, and capital investments could then be justified.

The people of Bangladesh use their boats and rafts to transport food commodities.

VIII. Ocean Freight

"And whatever you do, whether in word or deed, do it all in the name of the Lord Jesus, giving thanks to God the Father through Him."

Colossians 3:17

The purpose for including this chapter in the *Logistics Handbook* is to give you, as the logistics specialist, a brief idea of what it takes for the bags of grain to reach you at your project warehouses. For the field logistics teams, it is not actually necessary for you to be concerned about all of this, but it is included in summary form here so that you might have a more complete picture. In most cases, you will not be required to be involved in this phase of delivery. It will all be handled by the donor through his assigned freight forwarder and his overseas agent. The responsibilities of the freight forwarder will be explained later in this chapter.

Most of the U.S.-donated Public Law 480 (PL480) food commodities handled by World Vision field project teams during the drought crisis in Africa originated from the Gulf Coast ports in the southern region of the United States. These commodities were agricultural products made available through the U.S. Agency for International Development (USAID) by the Department of Agriculture.

When a relief project has been approved and USAID commits a designated amount of food commodities to a project, World Vision Relief Organization (WVRO) appoints a freight forwarder to direct the total shipment from the point of origin all the way to your distribution center warehouse. WVRO will coordinate with USAID when selecting the freight forwarder.

WVRO will select only the most reputable and reliable freight forwarder, giving consideration to its cost of services. A stand-by freight forwarder used on many occasions by WVRO, which has proven its reliability and dependability, has been Daniel F. Young, Inc., with headquarters in New York City. This company has been in the freight forwarding business for over 30 years and is well known and well established in all the major foreign ports of entry throughout the world.

In 1985, WVRO could foresee other possible expanded services in handling PL480 commodities. As a result, they selected a company, Marine Overseas Services (MOS), to handle the 10,000 MT of maize bartered from the Ghana government for transport to World Vision's approved relief projects in Mali. MOS will be discussed in more detail in Chapter XIII, "USAID-Sponsored Programs."

Freight Forwarders

A foreign freight forwarder is licensed by the Federal Maritime Commission (FMC) after proving its qualifications and reliability. It is required to file a surety bond with the FMC to prove its financial soundness.

The freight forwarder, when under contract with WVRO, is responsible to direct the movement of commodities from the point of origin to final destination. The forwarder assumes the responsibility for the protection and security of these commodities until your warehouseman receipts them at your distribution centers. The freight forwarder is liable for any loss or damage of the commodities en route to final destination until the ownership passes to the field team at the distribution centers. This point of ownership transfer is when the bags are unloaded from the trucks and stacked in your warehouse.

When the contract is made with the freight forwarder, he receives a copy of the vendor's invoice when he begins to prepare the details of the shipment. He plans the detailed routing of each shipment depending on how the client (WVRO/WVI) specifies the rate or quantities of the order to be shipped.

■ *In one of our earlier relief projects in Mozambique in 1984, USAID allocated 10,000 MT of maize to be shipped from a Gulf Coast port to Harare,*

Zimbabwe, through the South African Port Elizabeth. After evaluating our capabilities of receiving this tonnage in Harare for subsequent distribution into Mozambique, WVRO/WVI specified that the freight forwarder ship 2,500 MT in four equal shipments, each shipment spaced about four to five weeks apart. D.F. Young, Inc., was contracted as the freight forwarder, which in turn designated its South African agent, Swift Exports, to receive the ships at Port Elizabeth, off-load the commodities into rail cars and transport the bags of maize by rail to Harare.

In the meantime, WVI representatives contracted the services of a trucking and warehouse firm in Harare to receive the train shipments. This firm, called T.A. Holdings, Ltd., was responsible to unload the rail cars and truck the grain to its warehouses. Subsequently, at WVI's direction, its trucks would then deliver to designated distribution centers for later delivery to villages in World Vision's area of responsibility in Mozambique. Most of the eventual deliveries to the villages were made by the provincial government trucks because of the rebel crisis in Mozambique.

The freight forwarder is required to ensure rigid control of the shipment. The freight forwarder's traffic personnel are assigned to the overseas areas to ensure coordination and proper handling. In any U.S. government-sponsored ocean shipment, it is required by law that 60 percent of the ships used in transport are American flag carriers. The freight forwarder complies with the law and obtains the lowest possible ocean freight rates which must be approved by the U.S. government.

When bookings are completed, USAID is advised of the name of the ocean carrier and the anticipated date of loading. The suppliers of the food commodities contracted by the U.S. government are given written booking notice showing the name of carrier, port of exit and other sailing information in order to coordinate movement from inland point in the United States to dockside.

The coordination of these shipments must be closely controlled by the freight forwarder, mainly to avoid any unnecessary demurrage payments to transporters. Demurrage charges are assessed to the shipper when transporters (i.e., trucks, rail cars, ocean vessels, etc.) must wait to receive the shipment. Demurrage charges are normally quite costly, and depending on who is at fault, the freight forwarder usually must cover these charges, since he is responsible to coordinate all movement of commodities. As a result, the freight forwarder ensures close coordination of all shipments to avoid these demurrage charges. Be assured that the freight forwarder will cover and protect himself so he will not be responsible for the demurrage charges. He will look for the agency which caused the delay, which in turn will be assessed these charges. Consequently, do not let your logistics team cause these delays and charges. If it is our fault, we will have to pay. Demurrage is an unnecessary expense normally resulting from improper planning and implementation.

The freight forwarder is also responsible to ensure that proper export documentation is prepared for the shipment. Having the proper issuance of shipping documents is an important part of the operation. The timeliness is particularly important, since the lack of timely and proper documentation could cause an unnecessary delay in the shipment. Proper documentation is also important to have when the shipment arrives at the overseas port of entry. Necessary customs clearances are a part of the documentation normally arranged for by the freight forwarder with the help of the U.S. government. The freight forwarder can also arrange for the necessary insurance to protect the shipper, WVRO, against loss or damage of the commodities in-transit.

When the ultimate destination of overseas shipments are landlocked countries, such as Mali, the freight forwarder is prepared to make all connecting arrangements using truck, rail or barge through his designated overseas agent. A reputable and reliable freight forwarder will normally have an equally capable and experienced receiving agent to make certain that your commodities arrive on schedule at your distribution center. Normally, all or a large share of the inland transport is paid for by World Vision. Consequently, economical transport must be contracted. The project team should closely monitor the freight forwarder's selection of the transporter and means of transport.

Port Operations

As the logistics specialist, you may never have occasion to be involved in port operations, particularly if your relief projects are located in a landlocked country such as Mali or Chad. However, if you are in a country with direct access to ocean freight shipping, you may very well become involved in port operations. The need to know something about port operations then becomes important. For our projects in Ethiopia and Sudan, the logistics personnel are involved in port operations in varying degrees. In the Ethiopia logistics organization, one of the sections is called Procurement and Port Clearances. It so happened that the individual who managed this section was experienced in both of these functions, and it was convenient for the logistics manager to combine them under this individual's supervision. In Sudan, the logistics team generally relies on its contractors, MALT (Management and Logistics Team, a World Food Program organization), to handle all the port clearances of food commodities that come into Port Sudan.

No matter how reliable you feel your contractor is in his support of your project, any contractor deserves and demands close supervision by our field personnel.

This is the reason our field logistics personnel should have a working knowledge of port operations. You should never rely totally and exclusively on your contractor without providing him some degree of supervision. This degree will depend on the experience, competency and attitude of the contractor's personnel. Do not overlook the fact that when the ownership of the food commodities passes to World Vision at the port, as has been the case in Ethiopia, you become responsible for the cargo. If anything should happen to the cargo from this point on, the donor holds us responsible and liable for all losses. We could file a claim against the contractor for his negligence, but this is not always successful. As the logistics specialist, there is no substitute for your supervision and control and knowledge of what is taking place at all times.

In late 1984, when the Port of Assab was processing thousands of tons of food aid and supplies into Ethiopia and was fast exceeding its capacity, we sent a logistics team into neighboring Djibouti to evaluate the port, rail and road capabilities for supporting Ethiopia through the Port of Djibouti.

A detailed analysis and report of findings were made of this survey. It included an analysis of the most competent and experienced freight forwarder doing business at the Port of Djibouti.

Specific recommendations were made concerning which freight forwarder should be used in the event that any of our food commodities had to be diverted away from the Port of Assab. The concern was that too many ships would be arriving at Assab in a short time frame, and with the limited unloading capacity of the port, some ships would need to sit at anchor in the harbor before they could be unloaded. This type of delay results in costly demurrage charges. To avoid these charges and the delays of processing the cargo, the ships could be diverted to the Port of Djibouti. At that time, this port was operating at only about 30 percent of capacity. In the survey it was found that the food commodities could be transhipped on to Ethiopia from Djibouti via the rail line and/or the highway linking the two countries. It was not until several months later that food commodities belonging to World Vision did in fact have to be diverted to Djibouti. Our contractor in Ethiopia selected the freight forwarder in Djibouti to receive, store and transship the cargo. This particular freight forwarder was not one of those considered

At the Port of Assab (Ethiopia), laborers unload wheat flour from an ocean vessel. The food was a gift-in-kind donated to World Vision by the Canadian government.

or recommended in the previous logistics survey. The agent selected did not perform effectively. As a result, some of the shipment was neglected and allowed to get wet from the rain, which caused considerable spoilage and waste. Since we had accepted ownership at the port, we were held responsible for the loss.

This is an example of where all too often we become guilty of relying too heavily on our contractor without proper supervision from one of our team members. Then when a situation like this backfires, it generally ends up costing unnecessary funds which we can ill-afford to pay. We have an obligation to our donors to protect and wisely spend their contributions that make our ministry possible. Stay on top of the situation. There is no substitute for close monitorship and supervision over all logistics activities that fall under the logistics umbrella of responsibility.

The unloading of cargo from a large ocean freighter at a major port is quite interesting to observe. Most of the food shipments are bagged at the port of origin in 50 kg (approximately 100 pounds) bags. Because of the large quantities of food aid shipped by sea to Ethiopia and Sudan, much of it was shipped in bulk and bagged at the port. A considerable amount of this bulk grain was spilled and wasted, particularly at Port Sudan because of the speed of handling and the unprofessional manner in which the bagging was handled. The grain was bagged at the port by local laborers who were paid according to the number of bags they filled and stitched. Without proper supervision, the laborers were careless and too quick in stitching up the bags. In some cases, only three or four stitches at the most were made to secure the bag. Consequently, in the process of loading and unloading the trucks, much of this grain was spilled and wasted before it ever arrived at the distribution centers. Bulk shipments of grain are least desirable unless the port facility is properly and adequately equipped to handle the grain and bagging operations.

In the event that your logistics team does become involved in port operations, it is important for you to have some knowledge of the variety of costs included in receiving, storing and transhipping of the grain cargo at a port. Listed below are the major costs of port operations:

1. **Port Dues** - Firmly set by the port authorities on routine shipments. Subject to waiver by the host government if the cargo is emergency food aid.
2. **Freight Forwarder Fees** - Includes full responsibility for cargo and documentation as it arrives at the port until loaded on forwarding transport. Subject to reduction if the agent is contracted over a longer period.
3. **Stevedoring Out** - Discharge of cargo from ship's hatches to dockside.
4. **Landing Charges** - Movement of cargo from dockside to the lay-down storage area.
5. **Local Transport** - Movement of cargo from storage area and loading on forwarding transport (based on length of haul).
6. **Storage Charges** - Normally free for the first 30 days.
7. **Surveillance/Superintendency** - Lump sum payment based on total tonnage. Includes inspection of cargo prior to discharge to fix responsibility for any damages en route.
8. **Port Dues on Transit Cargo** - Fee per ton of cargo imposed by port authorities. Subject to waiver or reduction by host government if the cargo is emergency food aid.
9. **Contingency Charges** - Rebagging, restitching, repairing, etc.

The freight forwarder or agent will normally invoice on the basis of 50 percent of the estimated charges to be paid before the arrival of the vessel. The balance is to be paid on completion of discharge. If the agent is contracted over a longer period of time, as compared to receiving only one vessel, his fees would probably be reduced by 35-40 percent as negotiated.

Proper documentation at the port is an important aspect of receiving ocean cargo. This is where a good agent who knows the host country requirements can be of great assistance and can save you much delay and expense in processing the cargo. Relief organizations will normally be exempt from paying any import/export taxes or license fees. Until the port clearance team in Ethiopia "learned the ropes," World Vision cargo was unnecessarily delayed at the Port of Assab waiting to receive the original bill of lading. Once the team learned that a simple letter of guarantee could release the cargo, the flow of cargo then began to run more smoothly. It was not necessary to wait on the bill of lading. Another example of where a good agent who "knows the ropes" is invaluable to the logistics team.

The key to a successful ocean shipment is obtaining a dependable and reliable freight forwarder. His fees should be reasonably competitive. As mentioned, relief supplies are shipped through the intermediary of such an agent. He deals with port and customs authorities and local agents regarding all transactions, formalities and costs connected with the handling of such goods. He also handles the insurance of the shipment and the forwarding of documents.

In the event the logistics specialist finds himself involved in any way with an ocean shipment, or it is agreed that World Vision will take possession of the food commodities at the port, the logistician should become well acquainted with the freight forwarder or his designated agent. There must be close coordination with the agent during the receipt and processing of the food commodities through the port until they are safely en route to your distribution center.

IX. Materials Management

"During the seven years of abundance, the land produced plentifully. Joseph collected all the food produced in those seven years of abundance in Egypt and stored it in the cities. In each city he put the food grown in the fields surrounding it. Joseph stored up huge quantities of grain, like the sand of the sea; it was so much that he stopped keeping records because it was beyond measure."

Genesis 41:47-49

Joseph was a godly man truly anointed by the Lord, but because he did not keep records of the grain that was in storage, he would have had difficulty qualifying as a World Vision logistician!

Of all the chapters in this handbook, this particular one, "Materials Management," is probably the most important to the logistics operation and to the logistics team. The sections, "Stock Control and Accountability" and "Storage," are the heart and soul of a relief project where valuable food commodities have been placed under the custodial control of the World Vision project team.

Materials management is the basis for any successful logistics operation. In addition to the stock control and accountability of the inventory, materials management also includes the storage, security, and proper disposal of the inventory. For food commodities, the disposal is the distribution to the hungry villagers. When we assume the responsibility as custodian of any type of relief goods, we also assume an obligation to the donor for complete accountability and protection of commodities. The donor may be the U.S. government, one of our own support countries, other sources who donate gifts-in-kind or the millions of people throughout the world who donate their money to World Vision. These donors expect us to be a proper custodian of the items placed under our control and supervision.

Many of our field teams have not known how to set up a stock control system, although some have done well by improvising.

In our project in Menaka in eastern Mali, a young Malian Christian, who had just graduated from a local college with a degree in accounting, was assigned to the logistics team. This young Malian, named Clement, used his initiative, and with his basic knowledge of accounting, improvised his own stock control system. In the absence of any formal training or instruction, Clement set up accountability records using a simple debit/credit system. The system was effective and served its purpose. In addition to organizing the accountability records, Clement also improvised the first covered storage in Menaka using available mud huts to receive and store the initial truck loads of maize which arrived from Ghana. Clement turned out to be a self-made logistics specialist.

In the other projects, World Vision has not been as fortunate in having such dedicated employees as Clement. Teams have not always maintained adequate accountability records, at least during the early stages of the projects. Later, as more qualified people were assigned to the teams (in the absence of any standard stock control system) records were improvised--which at least kept us out of any serious trouble. Fortunately, thus far, no major losses of commodities have occurred because of shortcomings in the area of stock control and accountability systems.

It was not until late in 1985 that a standard stock control system was introduced to the World Vision Ethiopia team. It is anticipated that this system, with appropriate modifications, will become a standard system for all future projects. Highlights of this system will be explained in this section, to include sample stock record cards and other forms used in the system.

When planning a project, the logistics specialist must ensure that the most qualified people are assigned to work the stock control system. This not only includes the records clerks responsible for maintaining the records, but also the warehousemen responsible for protecting, securing, and receiving and issuing the commodities into and out of the warehouse.

Effective accountability and security of the inventory must include the entire period from the time the commodities are received at the distribution center warehouses (or the U.S. port, as the case may be) until they are delivered to the villagers. Management control of the entire inventory is a must if losses are to be avoided. This control must be as professional, under the circumstances, as it can possibly be.

Stock Control and Accountability

Stock control is the proper accounting for all the supplies, equipment and commodities in the inventory at all levels of the project. This accountability begins when the ownership of the commodities first passes to World Vision. In Ethiopia this happened to be at the Port of Assab when the commodities were first unloaded from the ship. In Mali, the accounting began when the commodities were delivered to the distribution centers in the Gao-Menaka-Ansango areas in the 7th Region, and in Nioro in the 1st Region. In Mozambique, the accounting began when the rail cars from Port Elizabeth (South Africa) were unloaded in Harare, Zimbabwe, and the bags of maize were stacked in our contractor's warehouse. At each of these receiving locations, the proper stock record cards had to be initiated as the accountability began.

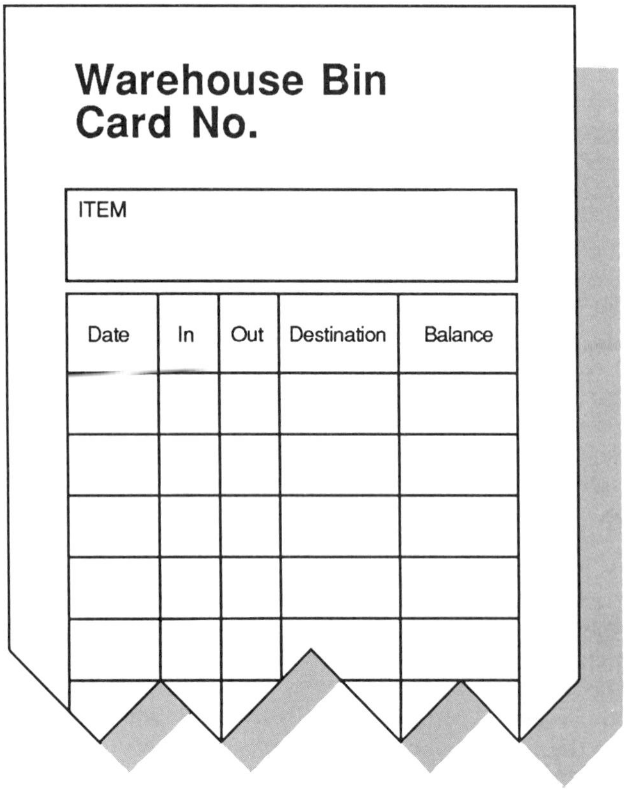

FIGURE IX.1

Single Stock Control System

The following describes the process for a simple one-warehouse stock control system. This system will work in almost any emergency relief operation or development project.

The logistics team warehouseman receives the first truckload of bags of grain. The warehouseman directs the unloading and specifies where and how he wants the bags stacked in his warehouse. As the bags are being unloaded and stacked in the warehouse, the warehouseman carefully checks each bag to note its condition. Any bags that are wet, or which have at one time been wet and have spoiled, will be removed from the good bags and stacked separately. Bags that are torn should be repaired, and the spilled grain rebagged. It is simple to tell if bags have been wet and dried. The grain will be caked and hardened, and it will usually have a bad odor.

Depending on the size of the warehouse, there will be several different stacks with a space of at least one meter between stacks. (A more detailed explanation on stacking will be given in the following section on "Storage.") As the warehouseman finishes a stack, he counts how many bags are in that stack and completes a warehouse Bin Card, attaching it to one of the bags on the upper corner of the stack (see Figure IX.1). This Bin Card is a running balance of the grain bags in each particular stack. The warehouseman notes the type of commodity in the stack, the date and the number of bags stacked on that date. Subsequently, as the bags are taken from this stack to be loaded on to trucks for distribution, the quantities, or number of bags, are noted by date and where they are going. A new balance of bags in the stack is noted on the card. A quick inventory in the warehouse can be taken by merely totaling the Bin Card balances on each stack.

Figure IX.2 is a sample warehouse Stock Card which is normally maintained by a warehouse clerk. This Stock Card is designed to account for the commodities from the time they arrive at the warehouse until they are loaded out for distribution. The card notes the type of commodity. Periodically, this card should be reconciled with the warehouse Bin Cards on the stacks in the warehouse as a cross-check for maintaining a current inventory of the commodities in the warehouse. From the balances on the warehouse Stock Cards, the warehouse clerk should be able to know at any time the current balance on hand, by commodity, in the warehouse. The column headed "Waybill No." is the number for that particular truck delivering the commodities to the warehouse or to the villages for distribution.

A waybill, or more formally referred to as a bill of lading, is a receipt for the goods evidencing their shipment. This document makes possible the delivery of the goods to the consignee. It is the basis by which the driver is paid when consignee has signed and receipted the waybill. For commodities being loaded for distribution to the villages, the driver is provided a waybill which is used to validate the delivery when it is signed at the receiving end. The bill of lading is an essential document used with truckers who are delivering commodities. A receipted waybill certifies that the goods have been delivered. For air shipments, the bill of lading is referred to as an air waybill.

Waybills are normally prepared and controlled by a responsible person who issues one to each trucker making a delivery from the warehouse. Each waybill is numbered for security and control. They should be kept in a safe place and accounted for. A standard waybill is usually printed in five copies, each in a different color which specify where each copy goes. The waybill will indicate the warehouse from which commodities are withdrawn and will note their destination. The waybill number is important to the warehouse Stock Card since it provides a cross-reference about where and when that particular truck load was shipped. Because it is a good record of each shipment, the warehouse clerk keeps copies of all incoming and outgoing waybills. Commodities should only be withdrawn from a warehouse when a valid and authentic waybill, issued by a responsible World Vision employee, is presented to the warehouseman. (The Commodity Management and Accounting Manual goes into further detail on the preparation and disposition of all colored copies of the waybill.)

The logistics specialist must insist that the warehousemen and warehouse clerks maintain accurate and up-to-date Stock Cards and Bin Cards. Since it is easy to forget to write down each entry, entries should be made at the time of the transaction. Timely entries will invariably result in more accurate records.

Even with two warehouses separated by several miles, such as in the Central Region of Sudan, the preceding simplified stock control system is sufficient. Each warehouse was supporting separate, unrelated distribution activities. Each had their own villages to support where monthly food deliveries were made.

Centralized Stock Control System

In contrast, a single stock control system would not have worked in a project such as in Ethiopia. There were as many as 15 camps being supported at the same time, out of a single depot complex--operated independently at each camp location. In this case, it was necessary to establish a centralized stock control system. This system is a bit more complex and requires people who know what they are doing and can direct the entire system to make it work. A centralized system was even-

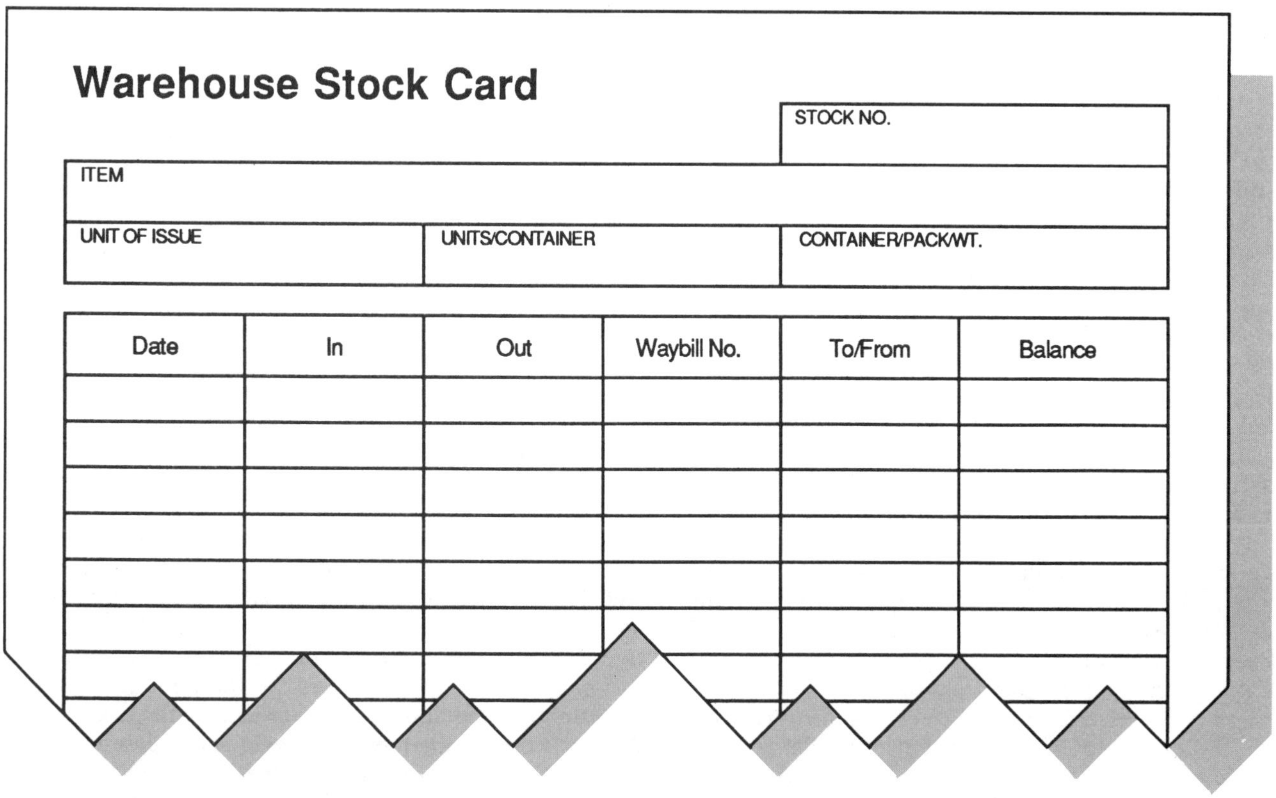
Warehouse Stock Card

		STOCK NO.
ITEM		
UNIT OF ISSUE	UNITS/CONTAINER	CONTAINER/PACK/WT.

Date	In	Out	Waybill No.	To/From	Balance

FIGURE IX.2

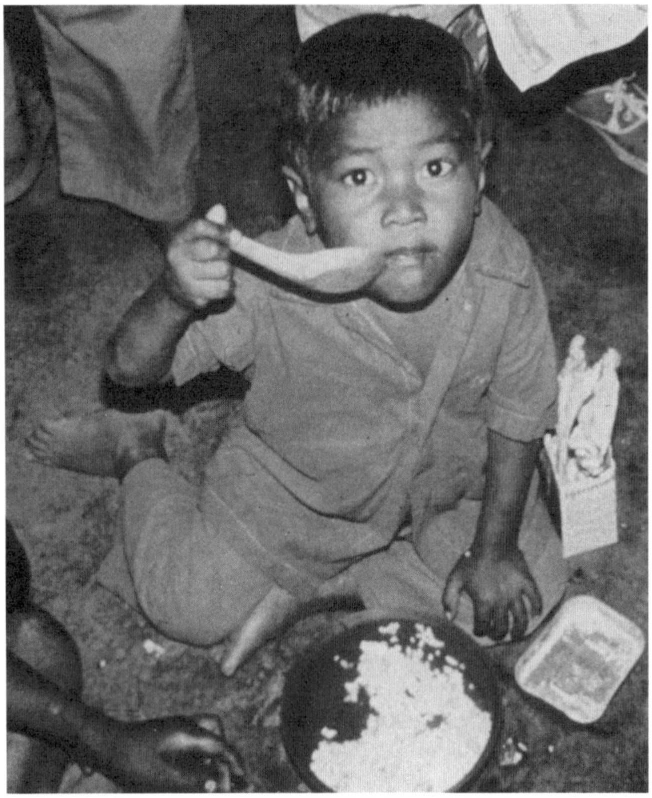

A Kampuchean boy receives food in one of the orphanages in Phnom Penh supported by World Vision.

tually established in Ethiopia. And once it was established, many of the problems in accountability were eliminated. The accountability system was greatly improved.

Certain things made this happen. First, an experienced systems expert from the International Office was sent to Ethiopia to set up the centralized stock control system. Second, local nationals with experience in accounting procedures were identified and hired to fill key positions in the system organization. Third, a complete reorganization of the logistics department was accomplished which made the stock control and accountability section a separate entity with the section chief reporting directly to the logistics manager. The reorganization gave this function the stature that it needed, and it made the staff aware of the importance of this particular function. Had these and other logistics actions taken place early in the buildup of World Vision's massive Ethiopia project, they would not have had so many growing pains.

The basis of a centralized system is a set of master stock record cards at a central location, in this case the World Vision Ethiopia headquarters in Addis. The stock record cards are simple 8" x 5" cards that are made up for each line item in the inventory. For example, there is one card for each type of food commodity being handled, one card for each type of medicine or clinical supplies being handled, one card for each item of non-expendable supplies being handled, etc. These cards are placed in a standard cardex file where the item description is visible to the records clerk. Because of the comparatively few line items being handled in the system, the daily entries were all made manually by the records clerk.

In a more complex inventory system that has many more line items, the system could easily be computerized. The manual stock record cards have the same information on them as the Bin Cards mentioned earlier (i.e., date, in, out and balance). There could be additional information included on the forms which would depend on the needs of the project. The main purpose of the centralized system is to be able to tell from these cards the current balances of each line item in the entire system including all the camps and depots.

To make this centralized system work effectively, the central stock record activity must be in complete control of the entire system. This means that the camps must submit their completed "Requisition/ Order" (R/O) forms to the central activity. Figure IX.3 is a sample copy of the R/O form used in the system in Ethiopia. This is the form used by the camps to replenish their supplies. Again the form can be modified to include any other information essential to the project. In Ethiopia, when the R/Os which originated from the camps arrived in Addis Ababa, they went straight to the manager of operations for his approval. Consideration was given to the operational needs, budget clearances and other factors requiring consideration by management. The R/Os were then passed to the stock control section which would make appropriate entries on the stock record cards and which would then direct the depot to ship the items to the particular camps. What if the depot does not have the item in stock that the camp requested? The central stock control activity will have this information because they should know the balances in the entire system--to include the depot. If the item is not on hand in the inventory, the stock control section notes on the copy of the R/O that the item is on "Back Order" and returns it to the camp. Then the stock control section notifies the procurement section that this particular item is not in stock and requests that additional items be purchased.

There may be some items that are so critical to the project that if a zero balance occurred and there were no items on hand, it could seriously jeopardize the project. It is a simple matter for the records clerk at the stock control section to prevent a zero balance situation from occurring. The records clerk can research this particular item and determine the frequency and quantity that the item is requested for by the camps. Then it is determined, with the help of the procurement section, how long it takes to order these items and how long it takes to receive them. This is called the "order and ship time." With this information, the records clerk can determine at what point or balance shown on the stock record card does a purchase order need to be initiated in order to have sufficient stock on hand to meet the routine requests from the camps. When the balance on the card reaches this level, a purchase order is immediately initiated.

It is important that the central records clerk know the balances in the entire system. The central office should require each camp to forward copies of their monthly inventories. When there is a zero balance on an item in the depot, it may be that the item requested may be on hand at another camp. The central records clerk will know this if he is provided copies of camp inventories. If the item is available in another camp warehouse, rather than noting a "Back Order" on the requisition, the stock control section will direct a shipment from the camp that has the item to the camp which has requested the item.

The stock control section is the overseer of the entire system. The depot and camps know only their own individual inventories, and the stock control section should know the balances at all locations. This centralized system will not work unless all players participate and comply with appropriate instructions from the stock control section.

■ *An example of unintentional non-compliance did occur in Ethiopia that resulted in a temporary breakdown in the centralized system. Some of the camps, to avoid delays, began sending their R/Os direct to the depot at Kombolcha. The depot filled these requisitions. Later, the depot sent a copy of the filled requisition to the stock control section--but too late to make the system work. This action did not allow the approval of the operations manager, nor did it conform to the centralized system. During this time the depot did not know their balances on hand. They were shipping items as they received an R/O from the camp. Once it was recognized what was happening, and appropriate management action was taken, the procedure got back on track and worked effectively.*

In a large operation such as in Ethiopia that has many separate camps and depots, the centralized stock control system is the most effective and the only way to go. The number of line items in the inventory will determine whether or not computerization is justified. In Ethiopia there were only about 600 line items, most of which were pharmaceutical items. In the event computers are not available, the manual system can be effective merely by adding additional stock record clerks, giving each a certain number of line items to handle. A manual system just takes longer to execute.

In a centralized stock control system, there must be a way for the stock control section to notify the depot to

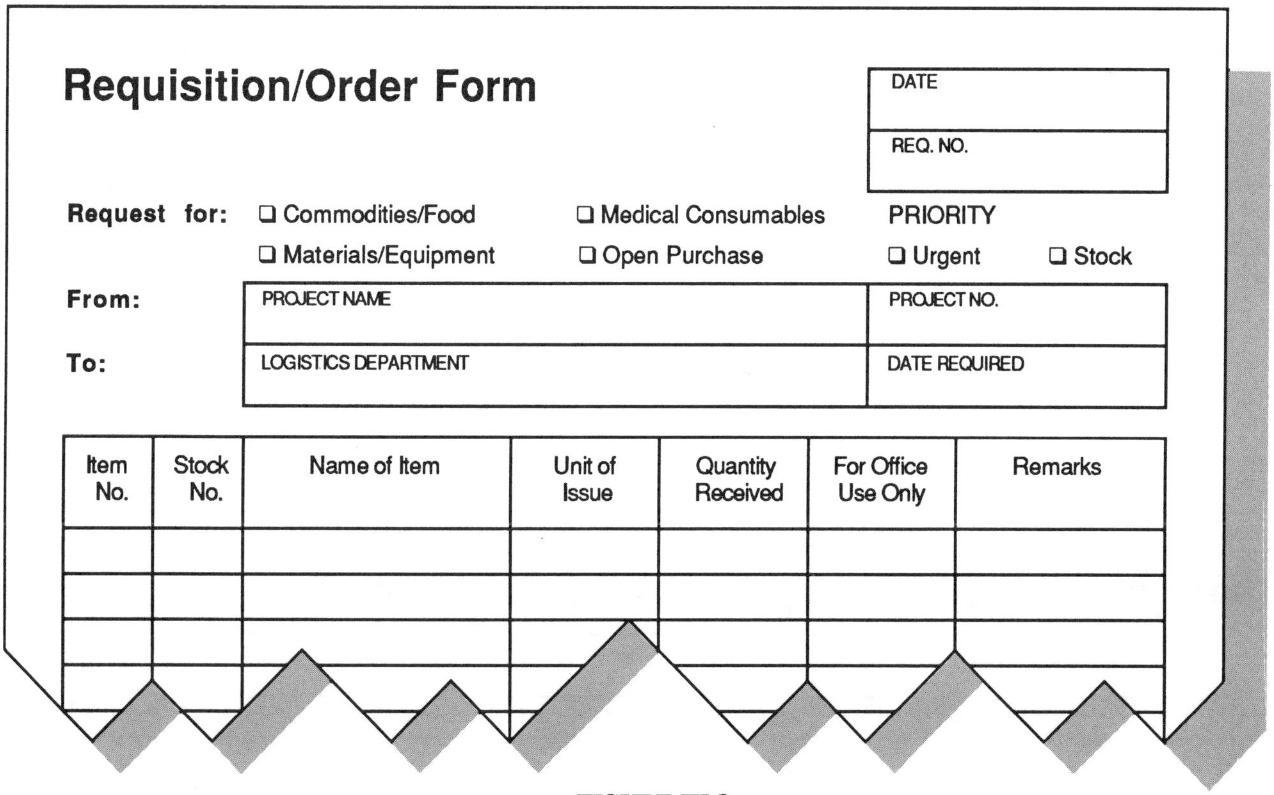

Requisition/Order Form

DATE

REQ. NO.

Request for: ❑ Commodities/Food ❑ Medical Consumables ❑ Materials/Equipment ❑ Open Purchase

PRIORITY ❑ Urgent ❑ Stock

From: PROJECT NAME | PROJECT NO.

To: LOGISTICS DEPARTMENT | DATE REQUIRED

Item No.	Stock No.	Name of Item	Unit of Issue	Quantity Received	For Office Use Only	Remarks

FIGURE IX.3

ship specific items to a particular camp. This is done by a completed "Shipping Request" form, a sample of which is shown on Figure IX.4. In the system used by WV Ethiopia, this Shipping Request form, with a copy of the R/O attached served as the bill of lading (waybill) for the truckers who were making deliveries to the camps. These shipping requests were sent to the depot by courier flying one of our two Twin Otters. Depending on the urgency and quantity of items requested, some orders were filled by the Twin Otters, but most were delivered by truck. Since the transportation was all owned by World Vision, and the drivers and crew were either employees or under contract with World Vision, the Shipping Request form could substitute for a bill of lading. A copy of the shipping request, duly receipted for by the camp warehouseman, was returned to the stock control section as confirmation that the items were delivered and received.

There is nothing difficult about setting up a stock accountability system. It is absolutely essential that some type of system be established so there can be a record of what was received and what was issued. The young Malian, Clement, used his initiative and his basic accounting training and set up a book to keep a record--a simple debit and credit system noting the quantity in and the quantity out. On this comparatively small project, this method was sufficient.

Presently, we are not using any standard system.

Shipping Request

❑ Urgent ❑ Routine

	DATE
SHIP TO	NAME OF PROJECT
MATERIAL FOR REQUISITION/ORDER NO.	❑ Copy Attached
DATE REQUISITION/ORDER RECEIVED IN WAREHOUSE	DATE FROM WAREHOUSE

Delivery to be made by:			
❑ Truck	VEHICLE NO.	DRIVER	
❑ Air	CARRIER		FLIGHT NO.

Shipping Information:

	Container Count	Weight ❑ Lbs. ❑ Kgs. ❑ MTs.	COMMENTS
Boxes			
Barrels			
Crates			
Bundles			
Bags			
Units			
Other________			
Totals			

SIGNATURE OF CARRIER	DATE	TIME
SIGNATURE OF RECEIVING PERSON	DATE	TIME

FIGURE IX.4

Logistics personnel attempted to go from Sudan to Ethiopia to learn their system. But time and work load did not allow this to occur. There is an advantage of having one standardized system. When World Vision logistics personnel are reassigned to other field projects, they do not need to learn a new system. It also makes the auditor's job much easier, and it is more economical since the same forms can be used on all projects, unless the project demands modifications to any of the forms.

It makes no difference what the items are that you must account for. They can be food commodities, medical supplies and equipment or materials and equipment used by the project team. Normally, consumable items such as office supplies, fuel, toilet paper and food are not formally accounted for. However, on other costly non-consumable items, sometimes called capital equipment, if you do not account for these items, you are vulnerable to becoming a victim of losses by theft or careless handling, as well as criticism from management and our donors. Loose accountability resulting in frequent losses could be disastrous for World Vision. It could jeopardize our relationship with the U.S. government, or any other major donor, on any future project. It is just good business to keep good records. This responsibility normally falls on the logistics specialist and his team.

It is anticipated that the commodity management procedures presently being introduced to our employees by Marine Overseas Services personnel in Sudan will serve to standardize stock control and accountability procedures among our project field teams. This is explained in Chapter XIII, "USAID-Sponsored Programs," and states that this current training program in Sudan will produce a Commodity Management and Accounting Manual which we can use in standardizing an accountability system. The purpose of the training program, and the preparation of the manual, will ensure that our procedures will be compatible with the reporting requirements of the U.S. government in handling PL480 commodities.

Storage

In emergency relief operations, the project field team will receive large tonnages of relief goods including food, medical supplies, blankets and other goods designed to bring relief to the depressed people. The logistics team must ensure that there are adequate facilities and space to store and protect these relief goods. In a relief operation of any size, there normally has been a sizable investment made in these materials. We have an obligation, as mentioned in the previous section, to protect and account for these goods. To do this properly and efficiently, there should be adequate covered storage or warehousing to store these goods until they can be distributed to the villages. Due to the urgency of the situation--particularly in remote areas--there will be occasions when there is no covered storage available, and the goods must be stored out in the open.

Open Storage

Most supplies and materials required in relief operations are vulnerable to loss by theft, pilferage, and spoilage when placed in open storage. This is particularly true for food commodities. As a result, open storage should be used only as a last resort. Open storage can temporarily work in an emergency, provided there are adequate means to protect and secure the commodities. Open storage should be considered only as a temporary measure to be replaced as soon as possible with improvised covered storage or warehousing. When the trucks begin to arrive with the goods, there must be a place to unload and store the goods. If no covered storage is available, you have no choice except to store the goods in the open. These trucks must be immediately unloaded and turned around, and some disposition must be made of the cargo.

This situation occurred in Menaka on the Mali project. In July 1985 rains had begun in the 7th Region of Mali. These were the first rains that had occurred in this area for over five years--God's answer to prayer! The truck haul bringing the maize to this region from Ghana was already en route. The first convoy of six trucks, each carrying about 35 MT of maize, arrived at a point some five miles short of the village of Menaka. The team had arranged for mud huts to be used as covered storage for the grain in the village and were ready to receive the trucks. Because of the heavy rains, a bridge over one of the run-offs (washes) that fills up during the rainy season had washed away. The trucks could not make it across to reach the storage facilities. The team had to immediately dispose of the 200 MT of maize, and the only alternative was to unload the trucks and store the maize in the open. Later, the wash was reinforced with rock sufficiently for a small pickup truck to go back and forth after the rains had subsided. When this occurred, the tedious job of shuttling the grain over to the mud huts began. The pickup truck would carry only about one-half of a ton, or about 10 bags. Particular care had to be taken because of the difficulty in getting the loaded pickup truck across the wash.

The logistics team came across from Menaka and selected the spot to store the 200 MT of maize which consisted of approximately 4,400 bags. Directing the unloading of the grain, the team selected a spot on the high ground to allow run-off of the rain. The soil was sandy which absorbed the rain water. The team was wise in not selecting a depression in the ground that would have collected the rain and caused many bags to become wet, which would have spoiled the grain.

The team collected all the tarpaulins from the

six trucks and used them to place on the ground underneath the bags and to cover the stack of bags as much as possible. With what was available in Menaka, there were sufficient tarps to protect the bags. These tarps were lashed down tightly with rope, and security guards were placed around the stack to prevent any pilferage of the grain. (Using the tarps on the ground underneath the bags is acceptable in most climates, especially the desert. Preferably, the bags should be on supports at least four inches off the ground. This is the height of a standard pallet which offers the best protection. It is also best to dig a drainage ditch around the stack with a run-off to prevent any water from getting into the stack.)

For quite some time during this rainy season, all of Menaka was isolated because of washed out roads and bridges. Follow-on truck convoys were held up in Ghana, and some were diverted to Bamako and Gao. The temporary open storage of the 200 MT near Menaka lasted only about two months. Not all of the grain had to be shuttled to Menaka. Eventually the team was able to get one of our Magirus 10-MT capacity trucks across the wash, and it was used to distribute maize to the villages on that side of the wash. The actions on the part of the logistics team to establish emergency open storage were proper and successful. None of the maize was lost or spoiled.

To summarize the criteria for the use of open storage, the following guidelines apply:

1. Use open storage as the absolute last resort, especially for vulnerable food commodities.
2. Select a location on the high ground which provides good drainage when it rains.
3. Use pallets or dunnage to keep the commodities at least four inches off the ground.
4. Provide adequate covered protection on the top and sides of the commodity containers. Securely lash down the protective covers.
5. Dig run-off drainage ditches around the stack to prevent any rain water from getting into the stack.
6. If possible, fence in the commodities and post security guards for 24-hour coverage around the site. If fencing is not possible, provide adequate security guards. Use roving guards at night.
7. Responsible members of the logistics team should frequently inspect the site to be alert to any pilferage or loss of the commodities.
8. Establish and enforce strict fire prevention measures around the site.
9. Obtain some type of covered storage (warehousing) as soon as possible and eliminate the need for open storage.
10. Do not detain the loaded trucks after they arrive at the project site. Get them unloaded immediately, or you will be paying the truck contractor a high demurrage charge for delaying the return of the trucks. (Demurrage: Additional penalty charge assessed to you as the consignee by the transport contractor for causing delays in loading or unloading cargo, thus tying up the contractor's equipment. These unbudgeted charges are costly and must be avoided.)
11. In making a site selection, in addition to selecting the high ground, provide some type of hard stand. A concrete slab is ideal, but alternatively using a gravel base or timbers offers much greater protection to the commodities. Sandy soil has comparatively good drainage.

Covered Storage (Warehousing)

Proper warehousing is the only sure way of adequately protecting, securing and accounting for relief goods and supplies. In WVI relief projects in Africa, many types of covered storage have been used ranging from mud huts and tents in Mali to school auditoriums in Sudan to corrugated steel buildings in Ethiopia. Some facilities are better than others, but all of them require varying degrees of maintenance and organization.

1. Security

The security of the facilities is a major factor in storing relief goods, particularly food. This was brought out quite clearly in late 1985 when the government of Sudan was marshalling tons of food commodities near the rail center of Kosti. These commodities were hauled from Port Sudan in long-haul trucks. At Kosti, they were unloaded and either warehoused, or transferred directly on to smaller contractor trucks for transshipment to the famine-stricken areas in the western provinces of Sudan. The marshalling yard at Kosti was enclosed by a wire fence. Armed gate guards protected the entrance to the yard and checked the trucks in and out of the yard. The thousands of hungry people in and around Kosti knew there was food at this marshalling yard. In their desperate state of hunger, many of the people would stand at the fence shouting and yelling for food while they banged on tin containers. These desperate actions resulted in food riots; several of these hungry people were shot and killed trying to forcibly take the food.

Starving people will be tempted to rob and kill to obtain food for their children and their families. This makes it a sensitive situation for World Vision to operate a warehouse complex in the midst of hungry people. Every attempt should be made to satisfy the needs of the nearby hungry people before beginning warehouse operations.

For these reasons, a warehouse is quite vulnerable to being broken into and requires adequate security measures at all times. The logistics team must be alert to these needs to ensure that warehouses are securely locked when not in use and further protected by security guards. If possible, the warehouse should be enclosed by a securi-

An Ethiopian loads bags of wheat flour from a World Vision warehouse.

ty fence with gate guards to control the access to the facility. The perimeter fencing around the warehouse should be well lighted at night. During the hours of darkness, there should be security guards who patrol the area. The extent of the security depends upon the conditions of the local area. Where there is a potential threat of food riots, like in Kosti, maximum security measures must be taken.

2. Fire protection

Fire is a treacherous thing to happen to a warehouse. A destructive fire could quickly bring a project to a complete stop. Considerable time would be required for the team to regroup and reestablish all the lost supplies and commodities, and to organize additional storage facilities. The best and only solution is to prevent a fire from ever starting. This requires good fire-prevention measures. Fire protection, like protection against theft and pilferage, is a major factor in planning for storage.

(a.) There must be equipment available to fight a fire should one get started. This is a major

Ethiopian families receive wheat flour donated by the Canadian government.

problem in the Third World countries of Africa where water is so scarce. Chemical fire extinguishers are a good investment; enough should be purchased to provide ample coverage of the warehouse.
(b.) Strict rules should be enforced which prohibit smoking or cooking fires in or near the building.
(c.) All employees should be well trained in fire prevention measures and should be instructed on how to use the fire extinguishers and all other fire-fighting equipment. Make certain an inspection schedule is set up for both the chemical and water extinguishers. These fire-fighting devices are not much good to you if they do not work when a fire breaks out.

3. Types of warehousing

Covered storage space is storage space within any roofed structure. The general-purpose warehouse, constructed with roof, sidewalls and walls, provides excellent storage for many kinds of items. There are many different types of special-purpose warehouses which field project teams do not need to be too concerned about. The only special purpose warehouse that could come into play in the field would be a humidity- controlled warehouse. This can be established by sealing off a portion of a general-purpose warehouse and using special equipment for controlling humidity. This type of a warehouse would be good to have and might be a necessity when storing certain kinds of drugs and pharmaceuticals. A refrigerated warehouse may be required when storing certain kinds of food or medicines. These types of special purpose warehouses require special provisions and equipment and, consequently, are not usually found in remote project areas.

A shed is a special-purpose warehouse which can be found in remote areas. This is a roofed structure without complete sidewalls and endwalls. A shed is used for storage of materials that require maximum ventilation or materials that do not require complete protection from the weather. This type of building is a compromise between open storage and warehouse storage. Tarpaulins could be used as sidewalls for the protection of supplies and commodities. Do not attempt to store combustible materials in a closed general-purpose warehouse. This includes any type of fuel or lubrication products, paints, filled propane tanks, etc.

4. Organizing a Warehouse

In relief operations, storage must be located on or near the main road and must be easily accessible to the larger long-haul trucks which will be delivering the commodities. There must be ample space for the trucks to unload. The warehouse should have sufficient doorways to simultaneously accommodate both incoming trucks and trucks being loaded out for distribution to the villages. These doors must be wide enough to allow the labor force to work.

Incoming commodities should be stacked by separate lot. Stacking commodities in the warehouse by lot provides better control for accounting purposes, and it allows compliance with a basic rule of warehouse operations--"first in, first out." Commodities that arrive in the warehouse first should be the first to go out to be distributed. This ensures maximum freshness of the commodities and prevents commodities from remaining in the warehouse over an extended period of time.

Even if a warehouse has a concrete floor or other hard stand, the food commodities should be stacked on wood pallets, timber beams or other types of acceptable dunnage. Whatever supports are used should be at least four inches high, which is about the height of a pallet. It is important to keep the commodity containers at this height off the floor or surface to prevent any moisture from entering the containers.

5. Insect Control

The floors of the warehouse should be swept clean before the stacking begins. It is important that the floors be kept clean of loose grain at all times.

One of our warehouses located at Wad Medani in Sudan was having a serious problem with rats and insects getting into the grain bags. Bags of grain which were being trucked from Port Sudan were carelessly and loosely sewn up at the port. As a result, considerable quantities of grain were spilled on the floor, and no attempt was made to keep it swept up. Since there were openings between the walls and the floor, it is understandable why they had the rat and insect problem. The loose grain was a clear invitation to the insects. Frequent cleaning schedules of the warehouse are the best way of keeping the pests and insects under control. Workers and staff should constantly be alert to rodents being kept under control.

In Mali the logistics team ordered two tent warehouses which were erected in Gao and Menaka. Particularly during the season of extreme heat and high humidity, these tents were a natural breeding ground for insects. It required frequent spraying and other insect control measures to keep the pests from taking over the warehouse. Insect control requires fumigation, and normally contractors with the right equipment are available locally to fumigate effectively. By closely inspecting a bag of grain in the stack, you can easily see if any insects have worked their way into the bag. You'll be able to see them feeding on the grain.

During the famine relief period in Kampuchea in 1980, one of the Communist countries contributed hundreds of tons of maize to the hungry people, There was one problem: the maize was totally infested with insects. The people, who were des-

World Vision employees in Wad Medani (Sudan) devised a means to feed malnourished children in our feeding centers. They used a cement mixer to mix the ingredients of a highly nutritious cereal.

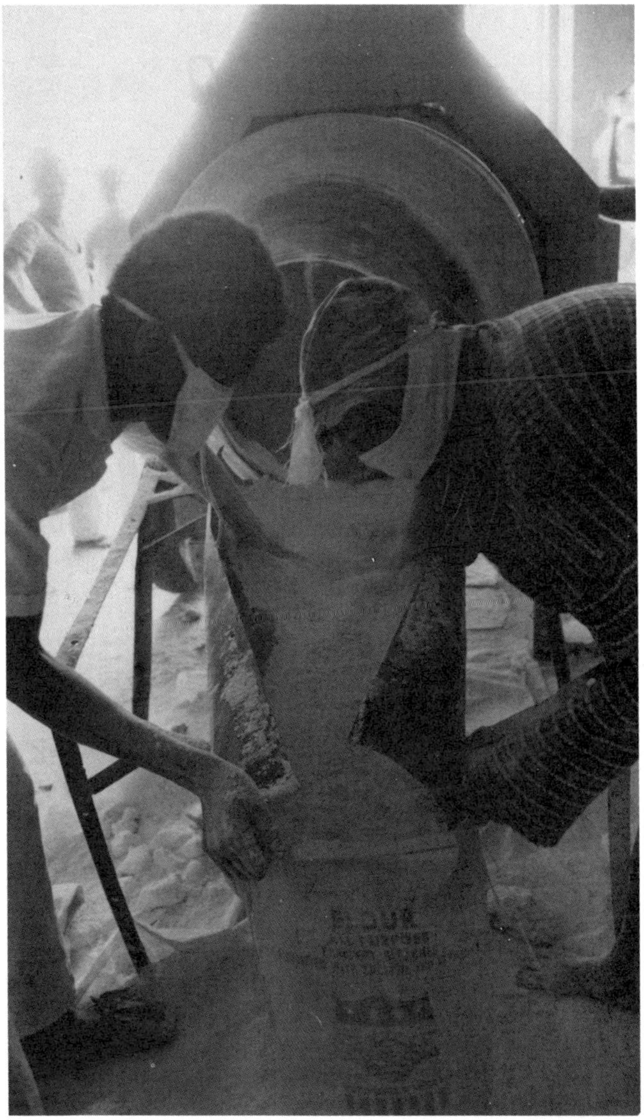

perately hungry ate the corn and, in so doing, became deathly ill. Many children died of severe diarrhea caused by the bug-infested maize. Even by long periods of boiling the corn, it was not possible to make it edible. World Vision, or any other agency, can quickly lose what good reputation it may have gained by distributing spoiled or insect infested grain to hungry villagers.

6. Warehouse Capacities

Two common questions frequently asked by field project teams: "How much warehousing is needed for the project?" and "How big should a warehouse be that will hold so many metric tons of bagged grain?" Warehouse space is easily calculated once you know how many tons you want to store. Once you know the maximum tonnage you will need to store, immediately plan to provide for sufficient warehousing to support the project. In remote areas you will not have the luxury of unlimited warehousing available. You must take what you can get and what you can afford. If the project is to receive a total of 8,000 MT of commodities, this does not mean that you must have 8,000 MT of warehousing. Distribution to the villages should be concurrent with receiving truckload deliveries. Just make certain that you have some means of communication to cut off deliveries in the event you encounter interruption in the distribution--heavy rains, truck breakdowns, etc. If you fill your available warehousing, you have no other choice except to resort to open storage for the overflow. This should be avoided if at all possible.

Warehouse space is normally calculated as a measurement of length times width times height (LxWxH) measured only to the eaves. Try not to store commodities above the eaves. They become unmanageable and too difficult to handle if they are stored too high. Remember these bags are being stacked by laborers, and the bags are heavy.

A warehouse should never be filled completely full using all available space; there must be space allowed between stacks for working aisles and ventilation and space allowed between the stacks and the walls. This space--normally about one meter--is essential when working a warehouse. The fire marshal also requires it. It is common in warehouses to paint lines on the floor to lay out stack boundaries.

The following sample calculation to derive the size of a warehouse building makes the assumption that the height of the stack was only two meters. The stacks can be higher than this but should not be so high that the bags cannot be conveniently handled by the labor force. Another consideration is that if they are stacked too high, the weight of the containers will rupture or burst open the containers on the bottom. Use good judgment when determining the height of the stack by considering the type of commodities being stacked. Grain bags could go higher than tins of vegetable oil. When stacking oil cartons or tins, make certain they are stacked right side up--a spill of vegetable oil can make a big mess!

Below are fairly accurate guidelines that can be used in determining warehouse storage space requirements.[1] (To convert cubic meters to cubic yards multiply by a factor of 1.308. Other conversion factors are listed in Appendix IV.)

One MT of:	***Occupies Approximately***
grain	2 m^3
medical supplies	3 m^3
blankets (compressed) (approx. 700 heavy blankets)	4 - 5 m^3
tents	4 - 5 m^3
blankets (loose)	9 m^3

The following is a sample calculation that may be helpful in determining the size of a warehouse building needed for a specific project:

"Assuming storage to a height of two meters, the minimum surface area occupied by the goods will be half their volume. Increase this surface area by at least 20% to allow for access and ventilation. For example, the approximate size of a warehouse to hold two months' supply of bagged grain for 30,000 refugees receiving an individual ration of 350 grams/day would be calculated as follows:

350 g x 30,000 x 60 days = 630 MT x 2 = 1260 m^3. 1260m^3 stored to a height of 2 m gives a surface area of 630 m^2. Add 20% for aisles and ventilation = 756 m of floor space. A warehouse building approximately 50 m long by 15 m wide would be required."

7. Materials Handling Equipment (MHE)

Normally field logistics teams need not worry about materials handling equipment. MHE would include specialized types of equipment used to work a warehouse--forklift trucks, dollies, conveyors, etc. In all of our relief projects in Africa, an inexpensive labor force has always been available and glad to work, sometimes for money but usually for food. Where ample manpower is available, MHE is not needed. If any MHE is available, it should be used for general purpose warehouses since many more commodities can be handled in a shorter period of time which gets the food to the hungry people faster.

The laborers are often weak from hunger and not always strong enough to handle heavy commodity containers.

An example of this occurred when we first arrived at the Zimbabwe-Mozambique border post at Mucumbura with food. The Mozambicans across the border in Tete Province did not have food, and they were weak from hunger. Arrangements were made with the village chief and border police to use a Mozambican labor force to unload the trucks on the Zimbabwe side. The first 75-100 workers who arrived to work were so weak that they were not able to handle the 100-pound bags. Between our team and the village chief, a plan was conceived to first have a big feast using the maize that had just been delivered. The women folk came over to do the cooking, and that evening a big feast was held. This was the first food of such large quantities that these people had seen in a long time. The feast was successful. The next morning these men began to handle bags in such a manner that it was a joy to watch. They sang and danced as they unloaded the trucks. An occasion such as this makes the job of a relief logistician all worthwhile!

The proper storage and handling of relief commodities will spell the success of a logistics operation. On a relief project there is usually no such thing as long-term storage of food commodities. The objective is to get these commodities into the hands and mouths of the hungry villagers as quickly as possible. However, storage is required as a temporary expedient for the purposes of setting up a well-managed distribution system, and providing a means to properly and accurately account for the commodities. The logistics specialist and his team must be on top of the storage function since it serves as the basis for all the other related functions to run smoothly and efficiently.

Checklist for Storage

The following is a convenient checklist that summarizes the highlights of this chapter.

1. After identifying project tonnages to be handled, select appropriate-sized buildings for warehouse space. Use as a guideline 1 MT requiring 2 m^3 of space for bagged grain.
2. Select a location on the main roads which can conveniently be serviced by large capacity long-haul trucks.
3. Make every attempt to get the host government to provide the warehouse building at no cost to World Vision, especially if it is owned by the national, province or local government. If payment is required, usually the first 30 days are free, and after that the cost is a nominal 2-4 cents per square foot per month.
4. Keep warehouse clean at all times.
5. Establish strict fire prevention measures and see that they are enforced by all personnel having access to the warehouse. Restrict the warehouse to only authorized personnel.
6. The staff and warehouse workers should be constantly alert to any possible insect, rodent or pest infestation. Immediate action should be taken to bring it under control when detected.
7. Use pallets, timber beams or other types of dunnage at least four inches high to keep the commodity containers off the warehouse floor.
8. Stack commodities by separate lots and use the general warehouse rule of "first in, first out" for all incoming and outgoing commodities.
9. Leave at least one meter of working aisle space between stacks and away from the walls for access, good ventilation and fire prevention.
10. Use good judgment when determining the height of stacks. Do not stack commodities above the

warehouse eaves for ease of handling. Be cautious not to stack commodities too high that will rupture or burst the containers on the bottom layers.

11. Provide adequate security of the warehouse. If possible, erect a perimeter fence around the warehouse and post security guards to control access to the warehouse. Also, provide perimeter lighting and walking night patrols if the situation warrants. Give high priority to feeding the local people around the warehouse.
12. Often, a vehicle maintenance shop must share available warehouse space. If so, erect a secure partition between the two facilities and enforce fire prevention measures in both facilities. Separate the two functions to the maximum extent possible.
13. The top priority for warehouse operations is the strict accountability of the supplies and commodities that come into and out of the warehouse. This is essential, and cannot be over-emphasized.
14. The warehouse should be dry and well ventilated. The roof should be in good repair, free from any leaks when it rains. The floor should be watertight, and free from any loose grain.
15. Identify any needs for special-purpose warehouses such as refrigerated, or humidity-controlled warehousing and determine availability.
16. Do not allow any combustible materials such as filled propane tanks or fuel to be stored in a closed general-purpose warehouse. Store these materials in a shed, or in the open.

---

[1]United Nations High Commissioner for Refugees, Handbook for Emergencies, Part One: Field Operations, Geneva, December 1982.

X. Maintenance

". . . Be strong, all you people of the land, declares the Lord, and work. For I am with you declares the Lord Almighty . . . and my Spirit remains among you. Do not fear."

Haggai 2:4-5

When World Vision elects to buy its own trucks, vehicles and other equipment, it inherits all the complex responsibilities that go along with the ownership. Some of these responsibilities include providing qualified drivers and operators, having on hand an adequate supply of spare parts, fuel, liability insurance, licensing and customs payments (unless exempt), and the most important responsibility of maintenance.

Preventive Maintenance

There is a sizeable investment in owning vehicles, but they are not worth much to you if they do not run. In order to keep vehicles running it is necessary to have qualified personnel, tool sets, equipment and a workshop to maintain them. In some circles, the accepted philosophy is to drive a vehicle until it breaks down and then worry about how to fix it. This is not the philosophy of World Vision. It is the responsibility of the logistics manager to ensure that a work-able, scheduled maintenance program is set up, and that all of these other requirements like mechanics, tool sets, shop space, etc., are available to get the job done.

In its simplest terms, the purpose of a preventive maintenance program is to properly care for the vehicles so they will remain in continuous service. It is a common attitude of operators, camp managers and others who have assigned vehicles not to want to give up their vehicles to the maintenance team for routine servicing, but this discipline must be established and maximum cooperation given to the maintenance team who can keep the vehicles on the road. The purpose of preventive maintenance is twofold. It is aimed at keeping equipment fully operable during periods of use. Then from a logistics point of view, its purpose is to prolong the life of the equipment, parts and components. Proper preventive maintenance practices can reduce costly, complex repair and can avoid prolonged deadline of the vehicles.

Field conditions often complicate the performance of preventive maintenance services. In the "bush" areas such as Menaka in Mali, it was not possible to find an enclosed structure that could be set up as a maintenance shop. Consequently, all maintenance had to be performed in the open. The blowing sand in the windy desert and exposure to the hot sun are not conducive to the enforcement of good maintenance procedures. In comparison, at Wad Medani in Sudan, the logistics team was able to rent a very substantial warehouse building with a concrete floor. A vehicle maintenance facility was set up at one end of the building. It was necessary for security reasons to install a wire partition to separate the maintenance shop from the storage of food commodities.

Even the best preventive maintenance program will not eliminate all equipment failures. In many countries where we are conducting relief operations, the terrain, weather and the demands on the vehicles place an unusual amount of wear and tear on the vehicles and their component parts.

The first step in establishing a good preventive maintenance program is to train the drivers on what to do and what to look for. A careless driver will normally be careless in taking care of his vehicle. The logistics specialist and his chief mechanic or maintenance chief should take great pains in training and supervising the drivers since they are the answer to limited downtime due to ineffective maintenance. This is not a simple task since most of the drivers are local nationals, and they are generally not inclined or knowledgeable in maintaining a large diesel truck or any other piece of equipment.

A diesel mechanic is a rather specialized individual. Even a driver of a diesel vehicle requires intensive training if he is expected to maintain it properly. The driver of a vehicle or the operator of a piece of equipment is the key to an effective maintenance program. If the drivers can be trained to take personal pride in the vehicles that have been assigned to them, and they are shown basic procedures for routine maintenance checks, the life of the vehicles will be prolonged, and there will be fewer vehicles in the shops for major repairs.

In one of our camps in Ethiopia, the vehicles were in very poor condition. The roads around the camp were rough and full of large rocks, but the drivers drove their vehicles like they were driving on a freeway. Management did not enforce any established driving disciplines because they themselves set a poor example of how to handle a vehicle. The excessive speed and abuse of these vehicles on the rough roads not only risked the lives of drivers and passengers, but the vehicles literally were being shaken apart, resulting in frequent deadline and short life of the vehicle. (This was also the camp where the management complained that they could not afford to be without a vehicle and therefore could not return them to the shop for scheduled maintenance.)

Well-trained and supervised drivers can extend the life of vehicles and prevent serious maintenance problems. Be alert to drivers who abuse vehicles and who are irresponsible behind the wheel. Establish a driver training and safety program, and conduct periodic refresher courses. Set up and enforce a hard and rigid policy involving careless driving.

Scheduled Maintenance

In addition to a preventive maintenance program largely built around the drivers, the logistics manager should ensure that his maintenance supervisor or chief shop mechanic set up a scheduled maintenance program for all World Vision-owned vehicles under his control. This is a routine servicing schedule when each vehicle, based upon time between services and mileage, is required to go into the shop. This schedule is no different than what an owner would set up for his own private automobile.

Engine oil and filters must be changed; the vehicle must be lubricated; and brakes, transmission oil, fan belts, hydraulic system and air filter should be routinely checked. The mechanic should be provided with a checklist for each type of vehicle, checking off and noting what he has done to the vehicle. A copy of this checklist should be a part of each vehicle's permanent file. During the routine maintenance checks, mechanics will note major discrepancies, usually pointed out by the driver who will describe the symptoms. When these discrepancies require major repairs, the field workshop should not attempt to perform such major repairs. Instead, the vehicle should be evacuated to a contract service facility, or to the dealer who should have competent mechanics, tools and equipment and component parts for replacement and repair.

Vehicle Workshop

To have an effective maintenance program, you must have an adequately equipped field workshop complete with tools and spare parts. The first requirement, of course, is to have a staff of trained mechanics and mechanics' helpers. The size of this work force will depend on the number and type of vehicles in the fleet. Before making any decision to buy a fleet of trucks and vehicles, you must thoroughly plan adequate workshop facilities. Facilities are essential for the protection of a major capital investment. In a crisis situation, logistics teams have a tendency to cut corners when there are not adequate facilities available. Sometimes this must be done as a matter of expediency, but unless the support facilities are ultimately made available, there is a risk of losing a large part of your investment.

Again, logisticians must improvise in "bush" areas where no better facilities exist. In Mali, when mechanics needed to be protected from the blowing dust and hot sun, canvas was used to provide a cover over the engine compartment, which allowed some protection. Without adequate facilities to work with, safety standards invariably are stretched to the danger point. Damaged tires that should be changed are left on vehicles. Engine oil is not changed when it is required. Trucks continue to operate with broken windshields.

The establishment of a workshop is expensive, especially if it is done correctly and it supports several vehicles and trucks of different makes and models. The major expense is in parts, tools, equipment and labor. A large spare parts inventory will require an accountable inventory control system, and an adequate security system to protect the investment. The replenishment of spare parts as they are consumed will require a procurement office with experienced personnel. Most spares will most likely need to be purchased and shipped from international sources. Customs clearances or exemptions will be required. A workshop, to be most effective, will require an available source of electricity and water. The diversity of makes and models of vehicles is a major factor in establishing a workshop. The more different makes and models of vehicles there are, the more complex the maintenance and workshop operations becomes.

When World Vision was considering a transportation contract to operate and maintain our fleet of trucks and administrative vehicles in Ethiopia, it was planned to utilize a "container workshop." This is a self-contained workshop that is shipped in four 20' containers, each containing various components of the workshop. The containers are placed in a rectangle preferably on a concrete slab 40' wide by 65' long. One container serves as an office, another as parts storage with bins, another is the electrical shop complete with generator, air compressor, drills, etc., and the fourth container is the workshop itself including four sets of mechanics tools. Also shipped in the containers are corrugated steel roofing, walls and double sliding doors for each end, making a compact, secure workshop built around the shipping containers.

In projects where vehicles are operating at widely dispersed camps where it would be impractical to return the vehicles to a central workshop for scheduled maintenance, consideration can be given to operating a mobile workshop. A compartment can be built and installed on the back of a 10 MT capacity truck. A driver and mechanic can move to and from the camps to perform routine servicing on the camp vehicles. A small number of various common spare parts can be included in the compartment, along with a supply of oils and greases. The capability of a mobile workshop is naturally limited compared to a fixed facility. Every two or three maintenance cycles the vehicles should be checked over more thoroughly in the base workshop.

Spare Parts

Spare parts are critical to any workshop operation. Repairs and servicing of vehicles cannot be performed without necessary spares. The maintenance manager must first determine what extent of repair he will be capable of performing on his vehicles. This will be determined by his facilities, quality of mechanics, available tools and special equipment and other factors. He can then initiate requisitioning procedures for the major spares he will need to keep his vehicles on the road. Tires have been one major item that have been used at a high rate in Africa relief projects. It has been difficult to keep enough tires in the inventory.

It is a common procedure that when new trucks are purchased, several months' supply of spare parts are purchased and shipped along with the trucks. This is not necessarily true for administrative vehicles, since the dealer is usually close by and is available to provide

This is Matt, a crippled Kampuchean boy who was found dragging himself on the ground until World Vision gave him clothes and a wheelchair. He witnessed his parents being tortured and executed by the Khmer Rouge. Behind Matt is his little brother who cared for him until they both were rescued and placed in Orphanage #2 in Phnom Penh.

spares and perform warranty maintenance as needed. In the case of trucks, they are usually too far removed in the field from the dealer or there is no dealer in the country that can supply spare parts.

Where the inventory of spare parts represents a large number of line items, the vehicle workshop must have a parts department and bin storage to keep the parts organized. A parts manager must be appointed to maintain accountability of these spares. He must anticipate the frequency for requisitioning certain spares to ensure that he does not run out of fast-moving parts. The parts storage is in a secure area which should only be accessible to the parts manager. The mechanic's tool sets are normally kept in the parts storage area and checked in and out each day by the responsible mechanic.

The one thing that can severely complicate the spare parts is to have several different makes and models of vehicles. In most cases, parts to different makes of vehicles are not interchangeable. Consequently, it becomes necessary to duplicate many spares, which compounds the storage and accountability problem. The logistics manager must be alert to this matter to make certain the project team does not purchase a variety of makes of vehicles. It has been a problem on certain projects when some team members like Toyotas, some prefer Nissans and others remain loyal to British Land Rovers. Unless the procurement of vehicles is controlled, it can seriously complicate the maintenance program.

Warranty Maintenance

The warranty is normally a part of the cost of the vehicle. But it is strongly advisable to purchase trucks and vehicles within your project country from a reputable dealer who can support your maintenance program. Certain major discrepancies occur on vehicles within the warranty period. If there is no dealer or his representative available to perform the warranty maintenance repairs, the warranty is not worth having.

In Mali for instance, there were no truck dealers represented in the country. As a result, our initial purchase were trucks that were made in West Germany--sophisticated Magirus trucks complete with a supply of spare parts. These trucks were air-conditioned, equipped with special accessories for desert operations. But they were a hydraulic nightmare. Spare parts that were not available had to be ordered from Germany, which took weeks for delivery. There were many occasions for warranty maintenance, but the capability did not exist.

A dealer with good maintenance and spare parts activities in-country can be invaluable to your maintenance program. Most of World Vision's Ethiopia truck fleet were the Dutch-made DAF vehicles. The DAF dealer in Addis Ababa was able to provide acceptable maintenance and spare parts support to our fleet.

Fuel

Another disadvantage to our owning our own fleet of vehicles is the responsibility for providing our own fuel. In the remote places, where there have been relief projects, this has provided a major logistics challenge. There are no service stations or easily accessible bulk distributors who are available to provide the fuel as you need it. Logistics teams have used various means of maintaining a supply of fuel in the "bush" areas. In the Gao-Menaka region, plastic tanks were used, each containing about 500 gallons of diesel fuel. In certain areas, it was necessary to bury these tanks to protect them from the hot sun and to provide a safer operation. In other areas, a protective cover was installed over the tank. On some projects, 55-gallon capacity drums were used, requiring a hand pump to refuel vehicles. Replenishment was infrequent by tankers who in many cases had to travel long distances. Or, our trucks would be used to exchange the drums. Fuel drums are a costly item. When handling fuel drums, plan to recover them for resale or reuse.

Refueling operations can also be a problem. Refueling requires additional people and takes time. It is much easier to lease the trucks and let the contractor be concerned about keeping his own vehicles maintained and refueled. He is normally better prepared and equipped to do this. Fuel of any kind is very sensitive to blowing dust, hot sun and theft. Active protective measures and security are important factors to consider when handling your own fuel.

Workshop Management

Setting up and operating a vehicle workshop requires experienced people and effective management. Maintenance records should be maintained on every vehicle. Each time a vehicle enters the shop for servicing or repairs, the vehicle is inspected by the shop foreman and a work order made out noting what is to be done to the vehicle. A copy of the work order becomes a permanent record of the vehicle and is included in the maintenance records.

A spare parts clerk is responsible to keep an accurate inventory of these parts. He is normally the one responsible to order replacement parts as they are needed, making certain he does not run out of fast-moving parts. If an essential part is not available when it is needed, the vehicle must be deadlined until the part is received, which denies the use of that vehicle to your most important food distribution mission.

Some shops get in the practice of "cannibalizing."

This is the practice of taking parts off of a deadlined vehicle to keep another running. Such a practice is not advisable simply because there is no end to the robbing of parts and you eventually end up with only bare frames of vehicles.

In some cases the replacement of parts has to come from international sources. This is a major challenge for a parts clerk since there is such a long order and shipping time to receive the replacement parts. Also, it requires a very knowledgeable parts man to be able to identify a part. A full set of parts catalogs and maintenance manuals should be available in the workshop. These are essential when ordering parts.

Operating a vehicle workshop of any size requires utilities, mainly electricity and water. When such a workshop is set up in the "bush," these utilities are not usually available. Power generators can sometimes substitute to operate electrical tools and equipment. Keeping these generators running is a problem in itself.

Other major tools and equipment which are important to a vehicle workshop would include air compressor, chain hoist, arc welders, battery chargers, grease pumps, oxygen and acetylene cylinders, jacks and many others.

It can readily be seen why we do not want the responsibilities that go along with owning our own vehicle fleet. Operating in the "bush," it is almost impossible to support vehicle fleets from a maintenance standpoint. These kinds of support activities greatly distract from World Vision's primary ministry of bringing relief and rehabilitation to disaster victims. It is much preferred to let a contractor handle all these ancillary problems.

Transport and Maintenance Contractor

It will, no doubt, be a long time before WVI will undertake a relief project the size and magnitude of what Ethiopia has been. World Vision responded to this crisis and made maximum use of the resources that were made available. Mistakes were made and lessons learned, but the job got done. As was pointed out in a previous chapter, we had no other choice but to get into the truck transportation business on a rather large scale. Consequently, due to the size of the project, World Vision ended up owning some 104 trucks and 109 administrative vehicles. Along with these vehicles, there were over 60 trailers. With this size fleet, all of these vehicles created a major operations and maintenance problem. As a possible solution to the problem, World Vision attempted to obtain a transportation contractor to handle these problems. A contract was actually entered into with a U.S. firm. The contractor sent nine professional expatriates to Addis Ababa. Work permits were not granted to these employees to allow them to go to the field and perform their job. Consequently, the contract ended up being cancelled and other arrangements made for us to handle these transport and maintenance matters ourselves. Had we been allowed to turn these functions over to a contractor, we would have been able to concentrate our efforts more on the actual relief operations.

The contract was planned for one year. The basis of the contract was a specified rate per metric ton of food hauled during this period. We knew approximately how many tons were to be hauled inside Ethiopia. With this specified rate, the contractor, using World Vision-owned trucks, would be responsible for the drivers of all trucks. The administrative vehicles were our responsibility except that the contractor was to establish a maintenance facility in Addis and set up a scheduled maintenance program. He was to handle the long-haul and short-haul distribution of all World Vision food commodities that came into the Port of Assab. He was to set up and operate a maintenance workshop in the depot complex at Kombolcha, being responsible for all the maintenance of the trucks and trailers. This would have been an ideal solution for our transport and maintenance problems for our large fleet of trucks. But it did not work out to our satisfaction. But this is an example of the type of contractor we should seek to have on large relief projects. In fact, it should go one step further to have the contractor furnish the trucks, if possible.

World Vision should avoid becoming distracted from its primary ministry by actually becoming actively involved in such a major transport operation. Logistics specialists are needed on such a relief project who are knowledgeable and experienced in these types of operations to monitor, direct and supervise such a contractor. WVI should avoid, whenever possible, getting into the trucking business.

XI. Other Functions

"Rescue the weak and needy; deliver them from the hand of the wicked."

Psalm 82:4

Logistics is designed to be flexible. In any type of field project, particularly in a remote area, there are many minor but essential support-type functions that must be handled as a part of the project. Usually, any activity related to the support of the project is eligible to be considered as a function of logistics. In most cases, however, any function that develops which the project manager does not know what to do with will be given to logistics--especially if the manager has an effective logistics team.

From the experiences gained during the drought crisis in Africa, this chapter deals with the "other" logistics functions that would be a logical part of the logistics team. There was not an effective logistics team in every project to handle two other logistics functions. Had there been, it would have been appropriate to assign these functions to logistics.

Construction Support

It is often necessary to arrange for the construction of facilities needed to conduct the relief operations. Such construction may involve erecting warehouses, maintenance workshops, fuel storage tanks, living quarters for the team or other similar type support facilities. Normally, construction work of this nature is contracted out to local contractors. Major construction, modification or repair is not within the project team's capability.

However, someone in the project team must take on the responsibility of coordinating this effort, and this would logically be the logistics team. Once the required action has been identified, the logistics team should draw up the specifications, locate a contractor and contract the work to be done. During the construction, some member of the logistics team will monitor the work to ensure that it meets our specifications. When the work has been completed, a final inspection has been made and the work is determined to meet acceptable standards, the logistics representative should see to it that the contractor has been paid for his work and the contract closed.

There are occasions when a job is not complicated enough to require a specialized contractor or construction materials. In such cases, the logistics team should take on the responsibility by hiring a group of laborers and direct their work.

In the Mali project, two canvas warehouses were purchased from Norway. The written instructions pertaining to the erection came with the kit. Even better, another agency operating nearby had recently purchased a similar warehouse. A technician from this agency was called on to supervise the erection of our warehouses. Local laborers were hired, specialized equipment rented and the warehouses were successfully erected at minimum costs.

A project team should identify any individual on the team who has had experience in any type of construction work. These people are invaluable to the team, particularly when the project is located in the "bush." This applies to other kinds of expertise as well, such as electrical, mechanical, etc.

Emergency Evacuation

One of the major concerns that a project manager would have for the welfare of his team, particularly operating in a remote area, is how to evacuate them to medical facilities if they should become injured or seriously ill. In many areas, the Lord has been with our field teams when there were no means of evacuation or even notification. The projects felt that they could not afford the "luxury" of a staff doctor, nurse or even a paramedic. However, in extreme remote locations where the team consists of several members, medical personnel are not a luxury, but a necessity. In these particular projects, the closest the team has come to having medical personnel are nutritionists who are normally working with the feeding centers. These nutritionists have been extremely helpful in treating minor cases, but they would not be expected to handle serious cases.

What is essential in a remote project is to have an emergency evacuation plan, and it should normally fall to the logistics team to draft up this plan for the project manager's review and approval. Any pieces of the puzzle that would complete an effective, workable evacuation plan should be provided, regardless of cost. The most critical item to an evacuation plan (which is usually missing) is some form of rapid communications. Radios are essential to any remote project, but they are often difficult and complicated to obtain. Usually the host governments tightly control the use of radios and the assignments of operating channels.

For a long time in Mali, we could not get approval for radios. The Mission Aviation Fellowship (MAF) came on the scene, which helped to relieve the situation. We were able to use the expertise of MAF, and we made certain to buy radios that had a frequency range compatible with MAF frequencies. MAF, through radio contact, was the team's answer to an emergency evacuation if it were ever required.

Planning for and conducting an emergency evacuation is a logical function for the logistics team. The plan must be well thought out, all project team members must be informed of the plan and, if necessary, able to rehearse the plan. When an emergency does occur, a person's life is usually at stake, and it is too late then to draw up a plan. With the rapid turnover of team members on a relief project, the evacuation plan should be in writing and should be a part of the orientation of any new arrival.

Radio Communications

The importance of having radios on a project was pointed out in the preceding section on Emergency Evacuation. But there are other good reasons for having radios on a project, particularly involving distribution of food. If possible, our trucks and drivers operating in these remote areas should be equipped with radios. Usually, these trucks are operating in the desert at considerable distances from the distribution center. En route to a remote village, a truck could break down, and without radio communications, there is no way of notifying a nearby truck or the base camp for assistance. The desert is no place to be stranded without necessary provisions for survival.

In Mauritania, our projects involved mobile teams conducting and supervising village feeding centers. These villages were miles apart, located in the desert. The Landcruiser vehicles were all equipped with radios that could communicate with the base station at our main office in Nouakchott.

The job of maintaining the project radios should be a function of the logistics team. It is very unlikely that the team would have a specialist who would know how to repair a radio or would have the necessary equipment if he did know how to repair it. However, the logistics team should designate one individual to coordinate the operations and maintenance of the radios. Find a repair facility in the host country, or perhaps use one of the other larger agencies which might have radio repair facilities, a supply of compatible parts and repairmen who are able to offer assistance. Work out the maintenance, repair and spare parts plans before the radios are purchased. Radios, like any other sensitive piece of equipment, require maintenance and repair. It is important to have well-trained operators who know how to use the radios and how to properly care for the radios when they are not being used. If this is handled properly, the radios will work when they are needed.

Of the relief projects conducted in the African countries where radios have been involved, the close coordination with the proper government offices of the host country has been an absolute necessity. Initial approval must be obtained, usually through a formal written request initiated by the project team. Since the governments are reluctant to approve these requests, continuous follow-up and persuasion is the order of the day until consent is given. If the logistics team is aggressive and has a good rapport with the government, this is another reason to assign this function to them.

Air Operations

In Ethiopia, World Vision owned and operated two Twin Otter aircraft. The air operations section that controlled these aircraft did not happen to be under the logistics manager. The project manager elected to keep this section under his own personal control. The aircraft hauled both cargo and staff personnel to and from the camps and feeding centers. Because of the close relationship to the logistics department, there was by necessity close coordination between the two staff agencies. Had this not been a personal choice of the project manager, the air operations section could have logically been placed under control of the logistics manager.

Air operations as a logistics function refers mainly to the air delivery or air-drop of emergency food commodities. As was mentioned in the transportation chapter of the handbook, the chartering of aircraft to deliver food is an extremely expensive operation. For an organization like World Vision to get into any large-scale air operations, the costs would be too high. However, it is not uncommon for governments to sponsor air operations, where food is either air delivered or air-dropped, depending on the terrain and location. In almost all situations, the aircraft and crews are military. Only the military forces are equipped, and their crews are trained to conduct air-drop operations.

When the decision has been made to conduct air operations and all necessary approvals have been received from the host country, the logistics team should be designated to coordinate the air operations. The assumption has been made that the food has been allocated to World Vision.

In the case of either type of air operation, it is necessary for a ground control team to be organized to receive the food after it is on the ground. The pallets of food must be cleared from the runway or the drop zone, stored and secured to be ready for distribution. The organization and coordination with the air crews is a normal function of a logistics team. Air operations is merely a special means of delivering and locating the food for distribution. Usually an air operation would not be justified unless the situation is critical and the need for the food is urgent.

The air crews will also place a request on the logistics team to provide assistance at the marshalling area to position and help load the aircraft.

Normally, the air crews will make arrangements to position a ground controller at the landing or drop zone. For an isolated area that is inaccessible to vehicle transport, these ground controllers and teams can be lifted into the area by helicopter. In Ethiopia, helicopters and crews from the Soviet Union were used in a similar type of air operation.

The use of aircraft is just another means for delivering the food commodities that will assist the logistics team in their function of distribution to the hungry.

Camp Life

The living conditions are an important factor in the spirit and morale of the project team. The mud huts that the Menaka team in Mali were required to live in at the beginning of the project were not conducive to high morale. With some initiative and ingenuity, the team installed a shower and a toilet. The acquisition of a small generator provided minimum power for a few electric lights and a small refrigerator. All of these basic "luxuries" were a tremendous boost to the team's morale. Thus we see that the logistics team, if comprised of talented and caring people, can greatly assist in improving these living conditions.

The logistics team can be assigned the responsibility of improving the living conditions. Included in these conditions are the feeding and kitchen facilities, recreation, convenient facilities to conduct daily devotions, etc. As was mentioned previously, the need for an individual who is handy with carpenter tools, electric wiring, mechanical installations and any other skills which may contribute to improved camp life is an invaluable person to have on the team. Such a person does not need to be an expatriate; he could be a local national.

The young people assigned to these teams need a diversion from their day-to-day chores of distributing food or feeding malnourished children in a feeding center. In Wad Medani in Sudan, all the team members were living under one roof in a fairly large rented, unfurnished house. They were in the field from early morn-

Kampuchean refugees arrive at a camp in Thailand.

ing until late in the evening, usually seven days a week. When they returned home and finished dinner there was nothing for them to do. By providing them with a few basic games, books, a TV and video machine, outdoor games, etc., there was a definite improvement in the morale and spirit of the team. The care and concern for the team members is a positive factor in their effectiveness and efficiency. Merely giving this job to the logistics team will usually cause things to happen, which in turn will improve the camp life.

Facilities Management

This is a fancy logistics term for a custodian. In most projects, the project team will be leasing facilities in the local communities such as team houses, workshops, warehouses, buildings for feeding centers and offices. Facilities of this nature are required to get the job done. When WVI enters into a contract agreement to rent or lease these facilities, the responsibility for the proper care and maintenance of these facilities should be assigned to the logistics team. When the project is completed, these facilities must be returned to the owner in the same or better condition than at the beginning. The logistics team can provide the liaison with the owner of the facilities. Often, improvements or modifications need to be made to fit the needs of our field team. The owner's approval and consent is usually needed before they are made.

Because of the logistics team's relationship to the overall support of the project, facilities management is another of the many additional responsibilities that can be given to them.

Care and Disposition of the Deceased

In the area near Gao during the Mali food distribution project, a contractor's truck engaged in hauling the maize from Ghana ran over and killed a laborer who had just finished unloading the cargo into a World Vision warehouse. The laborer, who was not a World Vision employee, fell from the back of the truck, and the truck's wheels crushed him, resulting in his immediate death. Since our team felt a sense of responsibility for this tragic accident, they immediately notified local civil authorities who made all necessary arrangements for the disposition of the body.

On fast-moving relief projects such as in Mali, accidents will occur. A serious automobile accident did occur in Ethiopia, requiring the urgent medical evacuation of the injured World Vision employees. Also, at one of the camps in Ethiopia, two local national World Vision technicians were killed by an attack from a group of rebels.

The project team must have a procedural plan of what to do when fatal accidents do occur. It is too late after it happens to decide what all must be done. This is another logical task for a versatile logistics team. Initially, it is necessary to coordinate with the appropriate branches of the local host country to determine exactly what is required from them. This is most important since you cannot act unilaterally when a fatality occurs. After a thorough investigation into all the requirements (including administration), the logistics representative should prepare a written procedure for the project manager's approval. The entire team should be familiar with the approved procedures so that any member will know what to do if such a situation does occur.

Insect Control

Attacking a major insect infestation is another example of how logistics must play a key role in relief projects. At the writing of this handbook, the locusts and grasshoppers are threatening the much needed harvest in sections of Mali and Mauritania. USAID in Mali reports that a "state of disaster" exists there due to an infestation of these insects. One official said that this year's infestation threatens to be the "worst agricultural disaster in West Africa in the past 50 years." USAID reports that as many as 250,000 hectares of cropland within a 400,000 square kilometer area are in immediate danger of being infested with migratory grasshoppers and locusts. Losses of 100,000 metric tons of millet and sorghum are possible.

The prevention of such a disaster will depend upon aerial spraying to prevent the migration. A major project of aerial spraying involves a challenging logistics operation. Wherever the spraying will take place, the support of the "crop duster" aircraft will be required. This involves a supply of fuel, storage facilities, and the capability of resupply. It will involve thousands of gallons of pesticides, which will also require storage and a means of resupply. Trucks and administrative vehicles, housing, maintenance workshops and other logistics facilities and resources will be required, not to mention a team of logistics specialists to manage and run the project.

This chapter has dealt with some of the other logistics functions that a project team could experience in the field. There could be others too numerous to mention. In development projects, there could be different types of functions, but common to both relief and development logistics functions, they must relate directly to the support of the project. After identifying major functions, the project manager and logistics specialist must evaluate and appraise the capabilities of the logistics team to take on these functions. A strong and aggressive logistics team can cause the project to run smoothly and efficiently.

XII. Personnel

". . . The harvest is plentiful, but the workers are few. Ask the Lord of the harvest, therefore, to send out workers into His harvest field."

Matthew 9:37-38

The effectiveness and success of the logistics operations of any relief or development project depend on the availability and qualifications of the personnel who are selected and hired to fill logistics positions. Once the logistics organization has been determined, based on the major functions that must be performed, then the challenge begins in selecting and hiring qualified individuals to fill the positions in the organization. WVI has experienced considerable difficulty in hiring qualified logistics specialists for field relief projects. Initially, the Personnel Department (HRD) recruiters did not know what qualifications to look for when recruiting for logistics positions. But WVI has since prepared a position description for recruiters to refer to. (A copy of this position description is shown as Appendix III in this handbook.)

The position description calls for a logistics specialist to be experienced in the major functions of logistics: namely, transportation, maintenance, procurement, materials management, including stock control, accountability and warehousing. He should be qualified to prepare and implement logistics plans in the support of relief and development projects. As a manager, he must be qualified to direct logistics operations involving the receipt, storage, accountability and distribution of relief commodities. Also, he must be capable of managing people, both expatriate and local nationals. Required to cooperate and coordinate with other activities of the project, he must also be a team player. And he must possess a certain amount of ingenuity.

Ingenuity is an unwritten trait of a successful logistician--conducting operations in the "bush" requires considerable improvising.

Ingenuity paid off in Mozambique. It was known that there were many hungry people in the extreme western part of Tete Province. Grain trucks were capable of moving food by improved roads through Zimbabwe from warehouses in Harare. Our food monitor in Tete Province knew that food could be delivered to the border, but the problem was getting the food across the border for distribution to the Mozambique villages. The border near the affected area was a rather wide, dry river bed with a sandy bottom. During the rainy season, this is a flowing river, becoming totally impassable. A physical reconnaissance was made to the border location by World Vision personnel. The Mozambique village chief with some of his staff came across the border to meet on the Zimbabwe side to discuss ways and means of getting this food across the dry river bed.

A plan of action was conceived, and steps were taken to implement the plan. After all the necessary government-to-government clearances were made to allow the commodities to be taken out of the country into Mozambique, the truck contractor was instructed to prepare three truck loads of grain. One of these trucks was to include a small farm tractor and trailer. While the trucks were loading in Harare and traveling the three days to the border, the Mozambicans, under the village chief, were busy cutting tree branches, lacing them together to place on top of the soft sand to provide traction for the tractor and trailer. The Mozambicans worked hard on this project since they were excited about receiving much needed food. The branches were an ingenious and effective solution for breaching this dry river bed when there was no other solution.

As it turned out, the tractor and trailer process was too slow. Greater quantities of food were needed in a shorter time period, and the rainy season was approaching. The governor of Tete Province agreed to make three 5-MT capacity trucks available for this project. The tree branch covering over the soft sand was reinforced sufficiently so that the trucks were able to drive back and forth, shuttling the food to covered storage on the Mozambique side as the distribution was being made to the hungry villagers.

In the business of conducting relief operations in the "bush," you will not always have the things needed to do the job, so you must be prepared to improvise by being ingenious. Where there is a will there is a way!

It has been very difficult finding individuals who have had experiences and meet the qualifications in all of the major functions of logistics. There have been individuals experienced in only transportation or warehousing or procurement, but to find a person with some degree of experience in all of them is rare--particularly in the civilian ranks. The ranks of the retired military have been a good source for logistics specialists. Normally, all the major military forces around the world, such as Britain, France and, of course, the United States, have produced individuals diversified in the field of logistics. World Vision has exploited the retired military organizations. There have been several prospective applicants from these sources with excellent qualifications, but job offers have been refused largely because of insufficient pay and benefits. The financial difficulties are too great when a person is asked to leave his home and family and go into a remote location for six months to a year. Not many professional logisticians have been willing to take on such an obligation. It has not been because of their lack of dedication or willingness to serve the Lord; they just have not been able to handle it financially.

In desperation, WVI has hired individuals who claimed to be logisticians but were not. In one case, a young man whose father owned a trucking company was hired as a logistician because he was from a trucking family. This did not qualify him as a logistician, and he failed in his job.

The six-month assignment periods have not proven productive. The person should be hired for at least one year. In six months' time he is just becoming familiar with the job when his assignment is finished. Also, the costs of rotating individuals every six months are high. In the case of logistics specialists, qualified individuals are too difficult to find to try to replace them every six months.

Professionalism and Christian commitment should be the primary criteria in recruiting. The project manager, on the recommendations of the logistics manager, should have the authority to dismiss incompetent individuals who are not capable of doing the job.

Local nationals are essential to a logistics team. They may be hired as mechanics' helpers, accountants, drivers, warehousemen, etc. Hiring nationals places the burden and responsibility of training on the expatriate logistics staff. The objective of the logistics team is to train the nationals sufficiently so that the

Here are satisfied faces of African children.

expatriate can eventually be released to return to his home country. Well-trained nationals can be a real asset to any logistics team. The logistics manager should emphasize the training aspect of the nationals assigned to the team. Training should be continuous on every project. And, training is not restricted to local nationals. With a continuous on-the-job training program, expatriates can improve their skills and learn new ones.

During times when World Vision experiences budget cuts and field personnel must be reduced, it becomes necessary to double up on personnel. Every effort should be made to keep the projects going because of their importance and the need to keep hungry people alive. If the food can be made available there should be a field team to distribute it.

Professional logisticians have generally had good management experience. A good logistician can manage people and resources. These qualifications would make him eligible to be considered for a position as project manager of almost any WVI relief project. This is the way a project operating on a reduced budget can double up on people. The logistics specialist, in addition to managing the project, can also direct the logistics functions provided he has a logistics team to carry out his instructions and to do the leg work. The project manager/logistics specialist can be the quarterback and call the plays. The objectives of the project can then be accomplished.

When budget conditions in World Vision become austere, which happens in most nonprofit humanitarian organizations, the emphasis on retaining qualified logisticians should be in the field teams. The greatest need for logisticians is in the field. Normally, the funding for organizations like World Vision is cyclical, and it is only a matter of time before the funding has been increased and the projects are once again accelerated. When this happens, it is best and wise to have a cadre of experienced professionals available in the organization.

In some of our relief projects, the project was implemented before adequate staff and essential resources were available and in place to handle the job. This has a significant impact on the few people who are there who must carry the load. It has been a major morale problem. And where there is an urgent need to save the lives of hungry men, women and children, the natural desire is to want to jump in before everything is ready. Of course, it is best to have the team organized on site, and to have all the vehicles and warehouses ready to receive and distribute the food. However, for an organization like World Vision, this is not the normal way to operate. Consequently, the pressure is put on the support offices and the International Office to expeditiously respond to meet the needs of the field.

Personnel is the key to any successful field project, and this is particularly true for the logistics field team. Unqualified personnel are a detriment to a project. The recruiters at all levels have a tremendous responsibility to provide timely support to the field teams. There are never enough people providing this backup support who have had field experience and who know what it is like to be stuck in a remote "bush" area without the necessary "tools" to do the job. The HRD staff should make frequent visits to the field in order to become completely knowledgeable in the field's personnel requirements. This is essential to providing adequate personnel support.

XIII. USAID-Sponsored Programs

". . . The troubles will soon be over, but the joys to come will last forever."

II Corinthians 4:18

The vast majority of food that World Vision distributed to help overcome the famine crisis in drought-stricken Africa since 1984 has been donated by the United States government (USAID). A total of 500,000 tons of emergency food was donated by USAID for Africa in 1984. In 1985, the amount increased to over 2 million tons.

The source of emergency food is largely from surplus and purchased agricultural products produced by the farmers of America. Authorization to grant these products is pursuant to the Agricultural Trade Development and Assistance Act of 1984 as amended by Public Law480 by the United States Congress. This is where the term PL480 originates. PL480 is mostly used when referring to food commodities donated and allocated by the U.S. Agency for International Development (USAID).

This congressional act authorizes USAID, a subordinate organization to the Department of State, to furnish agricultural commodities to meet famine or other urgent or extraordinary relief requirements; to combat malnutrition, especially in children; to promote economic and community development in friendly developing areas; and to promote feeding programs for needy persons, nonprofit schools and preschools outside the United States. Nonprofit voluntary organizations, like World Vision, which are registered with and approved by the Advisory Committee on Voluntary Foreign Aid are designated to receive these food commodities after submitting an acceptable operational plan. World Vision shares with the U.S. government a significant part of the funding necessary to cover the costs of transportation and all other related project costs.

PL480 Commodities

Title I:

Authorizes the President to negotiate and carry out agreements with friendly nations to provide for financing the sale of U.S. agricultural commodities on credit terms. The authority contains provisions to protect the growth of self-reliance within these countries and safeguards that such sales do not disrupt world prices of agricultural commodities. Also, precautions ensure that such sales do not displace normal sales of U.S. agricultural commodities.

Title II:

Authorizes the donation of food for developmental, nutritional and urgent or extraordinary relief purposes. Title II is primarily to supplement and to reinforce other developmental and nutritional activities within a framework of increasing local participation in management and funding. Title II is used when disasters strike in foreign countries, in overcoming malnutrition problems within vulnerable groups and for development work. In development work, emphases are given to maternal child health, food-for-work proposals and primary school feeding, in that order. Title II is viewed as interim assistance to enable countries to reach specific objectives to combat hunger, alleviate malnutrition, improve economic and social development and/or increase food production and improve its distribution.

Title III:

Permits eligible countries meeting specifically agreed conditions to sell commodities provided under Title I and to provide those local currencies against the dollar repayment obligation of that country.

The overall goal of Title III is to increase the access of the poor to a growing and improving food supply. Resources made available through Title III should be used for programs of agricultural development, rural development, nutrition, health services and population planning. Particular emphasis is placed on assisting small farmers, tenants, sharecroppers and landless agricultural laborers by expanding their access to the rural economy.

From the above descriptions of the various classifications of PL480 commodities, Title II best describes the food allocated to the drought countries of Africa where World Vision has conducted relief operations.

PL480 food commodities are normally grains (wheat, corn, rice and sorghum), processed grains (cornmeal, flours, etc.), vegetables (lentils and beans), blends (corn-soya milk, soy-fortified sorghum, grits, etc.), vegetable oil, dry milk, butter and cheese.

There is a separate piece of legislation referred to as Section 416, which is designed to enable the donation of dairy and other products to assist needy persons overseas and which is administered by USAID. Recently, World Vision has sponsored projects in Poland involving the donation of U.S. dairy products under the provision of Section 416.

Regulation 11

USAID has published strict guidelines and procedures for agencies to follow in the administration and handling of PL480 commodities. These procedures are called Regulation 11. Each agency is expected to be familiar with the provision of Regulation 11 and to comply with its procedures. One important section of this document is titled, "Records and Reporting Requirements." This section specifies that each agency handling PL480 commodities "shall maintain records and documents in a manner which will accurately reflect all transactions pertaining to the receipt, storage, distribution, sale and inspection of commodities. This shall include a periodic summary report and records of receipt and disbursement of any funds accruing from the operation of the program." When an emergency relief program under PL480 is initiated through a voluntary agency like World Vision, the Mission (usually the Food for Peace officer in the project country USAID office) will make a determination regarding the ability of the agency to perform the required record-keeping requirements outlined in Regulation 11. The purpose for maintaining the essential records and information is to account fully to the U.S. government for the donated commodities.

Regulation 11 specifically states that any relief agency handling PL480 commodities is subject to inspection and audit by U.S. government representatives enabling them to examine activities and records to ensure that they are being properly conducted and reported. At the discretion of USAID, all or any part of the assistance provided under the program can be terminated if the provisions of Regulation 11 are not strictly complied with. Any agency that has failed to comply may be suspended or disqualified from further participation in any distribution program under USAID assistance.

This plan must first be approved by the field director and the International Office; then presented simultaneously to the local USAID Mission and USAID/Washington,D.C. The field submits the plan to the local Mission and to Washington by WVI through WVRO. Regulation 11 specifies that the voluntary agencies have certain obligations to USAID in order to qualify as a participant in a relief project. One of these requirements is to submit for approval a Plan of Operation. This plan should include the following information:

1. A description of program goals and criteria for measuring progress toward reaching the goals.
2. A description of the target population, and the rationale for selection of the target group and selection of geographical areas where the program is to be carried out.
3. A logistics plan that demonstrates the adequacy of port facilities, transportation facilities and storage facilities to handle the flow of commodities to recipients without undue risk of spoilage or waste.
4. A plan of distribution to the recipients.
5. Statements involving the proposed cooperation and support from the host country and other participating agencies.
6. Description of the method to be used to supervise and monitor the distribution of the commodities.
7. Verification showing approval of the foreign government to import the donated commodities duty free.
8. Description and composition of rations and plans for evaluation.

Regulation 11 specifies that the voluntary agency (World Vision) must provide adequate supervisory personnel to ensure the efficient operations of the program. These personnel are to plan, organize, implement and evaluate the program involving the distribution of commodities. Maximum use of volunteer personnel is encouraged by USAID. It is specified that U.S. voluntary agencies shall be represented by a United States citizen, resident in the country of distribution, who is responsible to the voluntary agency for the distribution of the commodities.

Voluntary agencies are required under Regulation 11 to perform audits on a schedule mutually agreed to by USAID (Washington) and the agency. These audits are to be made by qualified individuals who are independent of those authorized to distribute the food commodities. The audits are designed to ascertain the effectiveness of the systems and procedures and to ensure that they meet the terms and conditions of the agreement.

The voluntary agency submits food commodity requirements to the USAID Mission showing the quantities of commodities required for each program. These quantities are submitted on a form specified by USAID.

It is the responsibility of the agency to determine that the recipients to whom commodities are distributed are eligible in accordance with the terms and conditions of the Food for Peace program agreement. Agencies shall distribute commodities only to eligible recipients without regard to nationality, race, color, sex or religious or political beliefs.

To the maximum extent practicable, and with the cooperation of the host government, adequate public recognitions shall be given in the press, by radio and other media that the commodities have been furnished by the people of the United States. USAID also expects appropriate recognition to be given at the agency's distribution and feeding centers. When the commodities are packaged for shipment from the United States, the bags are normally marked with appropriate identification showing the USAID emblem with the notation that the food is furnished by the people of the United States. World Vision's emblem is also normally shown. There is a notation on the bag that the contents are not to be sold or exchanged. Where we are required to reprocess or repackage the bags, they must, where practicable, remark the new containers. When bags are finally received by our logistics team at the distribution centers, many of the bags are torn or broken from all the handling, resulting in spilled grain. Much of this grain can be safely repackaged and used. However, in the "bush," it is seldom practicable to remark the new containers as specified by Regulation 11.

The field project team must ensure ahead of time that all food commodities shall be admitted by the host government duty free and exempt from all taxes. It is advisable to get this agreement in writing by a responsible branch of the host government.

It is common that a certain percentage of the shipment of food commodities arrive damaged or spoiled. For countries such as Ethiopia where the ownership of the cargo passes to World Vision at the receiving port, our responsibility is to advise the U.S. government representatives of the damages and spoilage. For landlocked countries, the procedure is somewhat different. USAID should arrange for an inspection by a public health official or other competent authority. If the commodity is determined to be unfit for human consumption, USAID will arrange for the disposition. These procedures are clearly spelled out in Regulation 11 and Handbook 9. When in doubt, consult your own commodity manager, or in his absence, the in-country USAID representative.

The logistics team should be alert to receiving damaged/spoiled grain at the distribution centers. Here the disposition of this grain is World Vision's responsibility. On several of the relief projects where truck transport has arrived with a load of grain bags, the individual truckers often have not adequately protected their cargos against rain. Either the tarpaulins have been inadequate or water has splashed up from underneath the

Food is portioned out to the people of Ethiopia.

A World Vision feeding center gives children of Sudan a chance to live.

cargo compartment causing some of the bags of food to get wet.

When grain becomes sufficiently wet, it will spoil and there is nothing that can be done except condemn the food for human consumption. At most distribution centers the locations are so remote that it is not possible to obtain an opinion of a public health official or any other competent authority. The decision must be made by the logistics team. You can only be expected to use your best judgment. When in doubt, the bags should be condemned. Spoiled grain which is allowed to get into the mouths of hungry people will make them deathly ill. World Vision's reputation is at stake if it allows this to happen. Condemned food must be isolated from the good food and the disposition should be clearly monitored and supervised by a member of the logistics team. Spoiled grain can often be used as livestock feed. But, the risk in turning spoiled grain over to hungry people to feed their livestock is that the people will be desperate enough to eat it rather than feed it to their animals.

USAID does allow a certain acceptable percentage of lost, damaged or spoiled commodities to occur in shipment. The quantities must be reasonable, or the loss will trigger an investigation by USAID. The logistics team must ensure that all such damages are properly and adequately documented, to include photos where appropriate.

If a loss, damage or spoilage of any part of the commodities is because of the voluntary agency's negligence--such as improper storage, care or handling--the agency is liable to the U.S. government for the value of the commodities spoiled.

The agency is responsible, according to Regulation 11, to notify the USAID Mission office of any losses. This notification, to be in writing, should include the circumstances surrounding the loss, damage or spoilage; any misuse; name of the responsible party; kind and quantities of commodities; size and type of containers; time and place of loss; current location of the commodity; and the estimated value of the commodity. There is no minimum dollar limit. However, the agency may elect not to file a claim if the loss or damage is less than $300. Often, the U.S. government changes the ground rules, so it is advisable to consult the commodity manager or the USAID representative.

All or any part of an assistance program sponsored by USAID may be terminated at any time at USAID's discretion. The undistributed portion of the food commodities reverts to U.S. government ownership.

The purpose of this section of the handbook is to give the logistics specialists an explanation of PL480 commodities and to highlight some of the provisions of the regulations which govern the control and accountability of the donated commodities. The logistics project team will be the ones most involved in the implementation of the USAID-sponsored programs and, therefore, should have a good understanding of what is required in their proper administration and handling. Before becoming involved in a PL480 commodity relief or development project, the logistics manager should obtain a copy of Regulation 11 and become familiar with its contents. Ignorance of regulations and procedures are no excuse to the U.S. government.

During the Africa drought crisis, World Vision became a well-respected customer agency of USAID. There were a few individuals in WVRO who were familiar enough with the U.S. government regulations that WVI was not allowed to jump into any project involving PL480 commodities without having the necessary experienced personnel to administer and handle these commodities in the field. WVI did not have any such experts with field experience. In order to satisfy USAID requirements of PL480 food commodities in Ethiopia, WVRO hired a young man who had previous experience in handling PL480 commodities. This arrangement worked out satisfactorily for all parties concerned. This WVRO employee became a part of the WVEthiopia logistics team and performed logistic functions in addition to his administrative responsibilities involving the accountability of the PL480 tonnages.

WVI's initiation of the USAID-sponsored program in Mali, soon after the Ethiopian crisis began, was a different story. To provide field personnel to assist WVI who were experienced in the administration of PL480 commodities, WVRO contracted the services of Marine Overseas Services, Inc., an international organization with their home office in Washington, D.C.

Marine Overseas Services (MOS)

The conduct of emergency relief operations using PL480 food commodities donated by USAID was a whole new experience for World Vision. It was recognized early on that people who were knowledgeable in the administration and accountability procedures of PL480 commodities were just not present in our organization. If we were to take advantage of the opportunity offered by USAID and become involved in the relief of drought victims in Africa, we needed this expertise.

MOS was a group built around the firm of Fettig and Donalty, who over the years have been in the maritime and foreign transportation business. MOS expanded its technical staff to include people who had many years' experience in handling USAID commodities. When the decision was made that MOS would become a contractor for WVRO, the officials of USAID strongly endorsed this marriage because they respected the reputation of the key management of MOS. This was a good start for World Vision.

The initial contract with MOS was to assist WVI in handling PL480 emergency food for Mali. Because of the urgent need for food in Mali and World Vision's desire to distribute it, USAID negotiated a barter arrangement with the government of Ghana which had a surplus supply of maize. The objective of the barter was to obtain maize quicker and closer to Mali in order to save the lives of the suffering people. The barter arrangement was successful, with the U.S. government agreeing to trade rice for the maize from Ghana. After a series of delays, the maize was on its way to Mali. MOS had arranged with a truck transporter to haul the maize to designated WVI distribution centers in Mali, to be met by the WVI logistics team.

In addition to arranging the truck transportation, MOS also agreed to furnish food monitors to assist WVI in monitoring the storage and ultimate distribution of the food. They also agreed to assist in the preparation and submission of all reports and records required by the U.S. government according to Regulation 11. The compliance with Regulation 11 was the main area where WVI benefited mostly from the assistance rendered by MOS. The stock control and inventory accountability of the food at the distribution centers, which is a key to submitting accurate reports to the satisfaction of USAID, was a new field for WVI. The MOS food monitors were helpful to WVI in setting up acceptable accountability records.

In mid-1985, WVI surveyed the famine crisis in Sudan and elected to become involved. This initial involvement was to assist the World Food Program (WFP) in the emergency food distribution in the Central Region of Sudan. Almost all of the food was PL480 food donated by the U.S. government. However, since the food allocation was made to WFP, it was their responsibility to meet the administrative requirements of Regulation 11.

WVI soon became disenchanted with working for WFP and sought their own program with the U.S. government. WVI applied to USAID and submitted an operational plan. USAID granted approval for WVI to receive a direct food allocation for distribution to the depressed people in the Central Region. Since we did not have the necessary experts to properly handle the administration of the PL480 commodities, we again through WVRO, called on MOS for assistance.

MOS had acquired the services of a man with many years of experience in handling PL480 commodities. This experience was both with USAID and with another Christian relief agency. MOS agreed to make this person available to WVI in Sudan. In February 1986, in a meeting with our project manager in Khartoum, the following specific objectives were agreed upon for MOS to accomplish during a specified period:

1. Design and document procedures to meet USAID/WVRO/WVI reporting standards for PL480 food commodities.
2. Implement system and administer the report of PL480 commodities distributed during period of contract.
3. Interface accountability system with logistics for compatibility with reporting procedures.
4. Provide on-the-job training for five selected personnel from the WVSudan team.
5. Turn over system to designated food commodity manager by May 15, 1986.
6. Prepare an outreach grant proposal in FY86 for WV Sudan for appropriate submission to USAID.
7. Train WVI staff to prepare a program plan for FY87, which includes Operational Plan and Annual Estimate of Requirements (AER) by March 15, 1986, as required by USAID.

In addition to the preceding objectives, MOS also agreed at the conclusion of the contract to prepare a written operating procedure, based on the WV Sudan program, but which would be applicable to any future WVI relief or development project involving the use of PL480 commodities.

With the concurrence of all parties, the program was implemented. However, it was soon to be interrupted when, due to a political scare in Sudan, all Americans were required to evacuate. After a few weeks, the staff returned to Khartoum and the program soon started up again. At the writing of this handbook, the objectives to be accomplished as previously listed are in progress. The suspense dates have been extended.

Food Commodity Management

The plan initiated in World Vision Sudan, utilizing the expertise of MOS, introduced a new discipline to World Vision which is planned to become an integral part of World Vision field teams in all future emergency relief operations. The plan is called Food Commodity Management (FCM).

The only logical alternative was for World Vision to train its own people. Sudan was selected to be the training ground utilizing MOS expertise. A small group of World Vision Sudan employees were selected to receive on-the-job training under the direction of MOS. This would provide a nucleus of our people who would be qualified to fill FCM positions in any future relief operations.

The FCM position is not intended to be a logistics position. However, many of the candidates who were selected to be trained in Sudan were out of the logistics team. The FCM and the logistics manager will be working side by side to accomplish the job. The logistics team will be the implementors for the FCM. They will maintain the accountability records from the warehouses and will be reporting distribution results. Many of the logistics records will serve as the basis for the FCM reports required by Regulation 11. Because of the reasonably high rate of turnover of our field personnel, the training of FCM personnel, like logistics personnel, must be continuous. The Commodity Management and Accounting Manual being prepared by MOS will become a working reference manual to complement this *Logistics Handbook*. Similarly, both publications will provide the continuity for field personnel. Key logistics specialists in the field are encouraged to become thoroughly familiar with the contents of the Commodity Management Manual. This is the only way to be effective in close relationships with the FCM. A mutual understanding of each other's responsibilities is required in order to be effective on the job. There is not expected to be any duplication in the two publications, or in the key personnel who will fill these positions. Each one will supplement the other, working in close harmony and coordination.

XIV. Contracts

*"Commit to the Lord whatever you do,
and your plans will succeed."*

Proverbs 16:3

On many occasions during the course of preparing and implementing a relief project, the logistics team will be faced with a decision whether or not to enter into a formal written agreement for goods or services. The purpose of this chapter is to provide you with the guidelines and policies of the International Office to assist you in acting as the representative of WVI in the field.

In most verbal or written agreements with another party to obtain goods or services, the dollar value of the goods or services is great. Consequently, the matter of entering into an agreement is a rather sensitive and critical issue that implicates not only your project team, but also the World Vision organization. As a result, act wisely, and do not be in a big hurry to get things moving. Make certain that all of your bases are properly covered before any agreement is consummated involving a financial commitment of WVI.

Verbal Agreements

Don't Move Too Quickly

In the haste of meeting emergency requirements during a crisis it is easy to make costly mistakes. Being an organization which relies on donor contributions, we cannot afford costly errors in judgment. Most mistakes are the result of wanting to move too quickly to get on with the project. Occasionally, in urgent situations field teams have taken the first available contractor, regardless of price. In most cases the agreements have been verbal just to get things moving.

A costly mistake occurred in the early days of our Mozambique relief project. World Vision had made an agreement with the Dutch government to distribute some 10,000 MT of white maize in Mozambique, which the Dutch purchased in neighboring Malawi. The target for distribution was Tete Province, which was an area severely hit by the drought. This province was between the countries of Malawi to the east and Zimbabwe to the west.

There were confirmed reports that many people in this area were dying of starvation and that malnutrition among the children was extremely serious. To make matters even more difficult, there was a civil war going on in Mozambique, and the rebels were operating throughout Tete Province. Some of the maize to be distributed in the province was being shipped by military-escorted truck convoys from Malawi, through the province, to Zimbabwe. World Vision was to arrange for a truck transporter in Harare, Zimbabwe to make further distribution from warehouses in Harare directly to storage locations and villages in Tete.

Our representative made a quick verbal agreement with the first and only truck transporter contacted. Like most truck transporters in Harare, this company had so much business that it did not need any more. But our representative was persistent in his demands to obtain this transporter's services to get the food moving. The transporter easily sensed his determination, and our representative accepted the first rate the transporter quoted. It was learned later that the accepted rate was nearly twice the standard rate being paid for similar services. The transporter felt that he was justified in his actions because of our representative's insistence that he take the job. Had our representative taken a little longer and made some phone calls to other transporters to obtain their quotes, he could have saved considerably in what was paid to distribute about 1,500 MT.

Formal Written Agreements

In determining whether a formal written agreement is required to purchase goods or contract services, consider the following questions:

1. **What is the amount of money involved?** The greater the financial commitment, the more important it is to have a written document setting out the provisions agreed to by both parties.
2. **For how long will the agreement be required?** If you know you have only 24 hours in which to locate and purchase a designated number of tents, blankets, etc., then the need for a formal agreement may be significantly less than if you have a period of time in which to negotiate and develop the contract terms. You should not, however, overlook the fact that a purchase order by telex can actually be a contract; care should still be taken in making quick decisions evidenced by telexes or verbal agreements by telephone.
3. **Is the subject of the proposed agreement already taken care of in another document?** If there is an immediate need to start up a health center because of an epidemic, there may be no need to begin formulating a written document with the local government if, for example, the field office's agreement with the government includes health care as one of the areas in which WVI may operate. Additionally, a purchase may be covered by a standard purchase order provided by the seller, which may only require your review of the contract terms to be sure they meet WVI's requirements.

If, after evaluating these considerations, you decide a formal agreement is required, you should then determine the following:

1. Who has the proper authority to approve and sign the agreement?
2. Have all supporting documents been made available to support a decision in favor of a formal agreement?
3. Who should prepare a draft of the agreement--the project director, a local attorney or WVI? After you have decided this with management, the packet of

Kampuchean orphans say "thank you" to World Vision in their own way.

supporting documentation (which should include the Project Planning Form (PPF) and an outline of preliminary items), should be forwarded to the appropriate individual. All of these should be translated into English if the originals are in the local language. It is your responsibility to obtain the appropriate approvals in writing from the International Office, either by telex or memo for your own protection as well as WVI's.

4. Have all appropriate steps been taken to finalize, execute and copy the agreement? Once a draft is completed, you should ensure that all details are correct and that all blanks are filled in. Additionally, you should when appropriate consult a local attorney to obtain a decision on the legality of its provisions under local law. Copies of the agreement should be sent to all concerned parties and you should ensure the safekeeping of the original. (NOTE: No matter who drafts or signs the agreement, a copy should always be sent to the International Office Legal Counsel and the individual having ultimate responsibility for approving the project, even if the agreement has already been executed.)
5. Are all contract terms being complied with? You should ensure that the terms and conditions of the contract are complied with and inform management if either WVI or the other party appears to be unable to perform its duties as described in the agreement. This advance warning enables WVI to take corrective action to reduce any damages that might result from a failure to complete the contract as originally agreed upon.

In most of our agreements, "time is of the essence," which means the dates involved are firmly set, and if a party cannot comply by the set date, the entire agreement may be threatened. You should attempt to ensure that agreed-upon dates and times are met. Any extensions or cancellations, as well as any other amendments to the agreement, need to be in writing and signed by the same parties who approved the contract originally. You should not rely on oral statements.

Legal Terms of a Contract

What is a contract?

It is an agreement between two or more persons intended to create a legal obligation between them and to be legally enforceable. Basically, and as far as you are concerned, there is no legal difference except in matters of procedure and proof between oral and written contracts.

The rules of different legal systems vary substantially on many matters. Generally, however, the law requires that there be "consideration."

What is "consideration?"

One party promises to perform in return for something offered by the other party. It is normally a payment, transfer of goods, doing of services or the surrender of another legal claim. Civil law systems generally can hold a contract enforceable even without consideration if an appropriate formality has been complied with.

What are the essentials of a valid contract?

1. Both or all parties must have had at the time of contracting, contractual capacity and legal power or authority to make the kind of contract contemplated,
2. Must have reached agreement on all the material terms of their bargain.
3. Must have intended a legally enforceable, as distinct from a merely social agreement, and
4. The agreement must not be objectionable on the grounds that it is intended to effect an impossibility or anything illegal or contrary to public policy or otherwise unenforceable.

What is "breach of contract?"

It is the failure by either party to implement any of the duties imposed on him or her by the contract and gives rise to claim of damages in compensation for the breach.

When may a contract be discharged and terminated?

By agreement, by due performance or payment, by waiver of performance, by replacement with new contract, by illegal performance, or by frustration when performance was different from what was originally contemplated.

It must be emphasized that obligating World Vision into any formal agreement, particularly where consideration is involved, is a sensitive matter and must not be taken lightly or done too hastily. Appendix II is a copy of WVI's policy on the "Execution of a Contract," which all field personnel should familiarize themselves with and understand.

A Successful Contract

In contrast to the previous example of how not to do it, following is an example of how a successful contract was implemented in a late relief project in Mozambique. This was a continuing project whereby USAID provided 10,000 MT of maize shipped in four separate increments of 2,500 MT each through the Port of Elizabeth in South Africa by rail to Harare, Zimbabwe for subsequent distribution into Mozambique. There was an ongoing emergency crisis in Mozambique, but the logistics team took the time to properly work out arrangements for a contractor to perform the services of warehousing and transporting the maize for WVI. WVI logistics representative went to Harare to make plans for the receipt of USAID-furnished maize. Having determined that a formal agreement would be required for these services, the representative first identified all the transport and warehouse contractors in Harare who were considered qualified to perform the work required. In order to keep it as simple as possible, the list of possible contractors was reduced to five firms which were considered after investigation to be the most reliable. Each firm was personally contacted to ascertain if it was interested in bidding for the contract. WVI's representative toured all of the firms' facilities to evaluate their capabilities for handling the job.

After the five contractors were selected, a Contract Specification was prepared. This was a one-page description of services WVI wished to have the contractor perform, which was mailed to the five bidders. The mailing included instructions that their bid responses were to be returned in a sealed envelope by a specified date.

When all bids were received and opened, they were analyzed as to costs, equipment and capability. A detailed evaluation and analysis were made of the bids and forwarded to I.O. management with recommendations. After management evaluated the bid responses, the contractor winning the award was notified.

Soon thereafter, the contractor met with the level of management who would be directly involved in the execution of the contract. The contractor was not the lowest bidder, but he was the most capable to perform all aspects of the contract. Since he was not the low bidder, the initial meeting resulted in some successful negotiations to reduce his rates.

Afterwards, both parties joined in drafting an appropriate contract. The contractor's legal staff reviewed the final draft, and a copy was sent to the I.O. where it was reviewed by management and World Vision's legal staff. Management officials from the I.O. came to Harare where a final meeting was conducted between the two parties. This meeting resulted in officials of both parties signing the contract. The contract was ready for implementation. The entire procedure as outlined took no more than three weeks to complete. It turned out to be a sound agreement, and it has continued throughout our relief projects in Mozambique supported from Zimbabwe. The contractor's performance has been most satisfactory. During the implementation of the contract there were minor modifications made to the contract. These modifications were made after mutual agreement between both parties and identified in a written document.

Included in Appendix I of the handbook is a copy of a sample contract agreement for services. This sample agreement has been prepared and provided by WVI's legal staff with the instructions that written contract agreements prepared in the field should follow this format with any pertinent revisions. The first part of the agreement is considered a standard "boiler plate" agreed to by both parties. This is followed by "Exhibit A: Professional Services," which actually sets out the exact services that are to be provided by the contractor. "Exhibit B: Fee for Professional Services" sets out the maximum amount due under the contract, and the conditions and method of payment.

Logistics field personnel who become involved in any formal agreement, verbal or written, implicating World Vision, should follow these procedures. When in doubt concerning the actions to be taken, request immediate advice and assistance from the International Office legal staff. This assistance should be requested before any contractual commitments have been made.

XV. Summary for Managers

"He has scattered abroad His gifts to the poor; His righteousness endures forever; His horn will be lifted high in honor."

Psalm 112:9

In the functional chapters of the handbook, the details of each major logistics function have been described to assist the logistics specialist in carrying out his duties. It is equally important for the project manager or field director to know what to look for and to be able to evaluate the progress of the logistics activities and the performance of the project logistician. The purpose of this final chapter is to summarize the major logistics functions and to highlight these areas as a guide and checklist for the project manager.

Project Planning Workbook (Chapter I)

1. After a project has been conceived, insist that your staff complete the workbook to the maximum extent possible. This should be done before the PPF is submitted.
2. The logistics specialist should concentrate his efforts on Section 10.0 of the workbook, "Logistics Planning."
3. This preliminary planning should identify any possible logistics constraints expected to be encountered during the implementation of the project.
4. The completed PPF itself should include the high lights of the preliminary logistics planning, which would assure the International Office that the logistics aspects of the project have been duly considered. A general narrative of an overall logistics concept would suffice for the PPF.

Logistics Plan (Chapter I)

1. Immediately after the logistics specialist has completed his portion of the workbook, he should begin the detailed preparation of the Logistics Plan.
2. A written plan should follow the outline of the major logistics functions as they appear in this handbook: General Concept, Organization for Logistics, Procurement, Transportation, Materials Management, Maintenance and Other Logistics Functions.
3. The purpose of a written Logistics Plan is to identify what must be done and what is needed to carry it out, to organize the logistics actions in an orderly manner, and to provide a checklist for implementing the project.
4. A detailed Logistics Plan should convince you, the project manager, that your logistics specialist is definitely on top of the logistics activities. After your review of the plan, you should be assured that all logistics have been well thought out and planned, and are being readied for implementation.
5. Appendix V is a sample outline of a Logistics Plan.

Organization for Logistics (Chapter III)

1. If a logistics specialist has not been assigned to your team, designate another "best qualified" individual to handle the logistics. Immediately submit a personnel requisition for a logistician to the International Office.
2. Spend ample time with your logistics specialist in the review of his written Logistics Plan. Question him thoroughly on any areas of the plan where he might seem vague or confused.

3. When satisfied with the plan, you and the logistics specialist should develop a proposed logistics organization. This should be based on the major logistics functions that must be performed.
4. Evaluate the existing personnel on your staff, both country nationals and expatriates, and assign the best qualified individuals to the key positions. Where critical shortages exist, do not hesitate to seek the assistance of the International Office and submit immediate personnel requisitions. If possible, anticipate these personnel requirements as early as possible. Give the HRD ample time to recruit.
5. The logistics specialist should report directly to you, the project manager. You should require periodic updates from him during the preparation and implementation phases of the project. He should be instructed to advise you immediately when problems arise. Request approval for the purchase of capital equipment and initiate requisitions as soon as possible.
6. Avoid too much fragmentation. Do not take on more than you have adequate personnel and resources to do the job effectively. If possible, have at least the key people and resources (trucks, vehicles, mechanics, drivers, warehouses/warehousemen, etc.) in place before the project is implemented.
7. In the event that food commodities to be handled are furnished by the U.S. government (PL480), ensure that you are assigned a commodity manager or others who are familiar with the appropriate USAID regulations and reporting procedures.

Procurement *(Chapter V)*

1. It is advisable to place the procurement function under the logistics specialist. Both goods and services will be procured in support of the project.
2. Obtain all valid field procurement policies and procedures disseminated by the International Office. In the absence of specific published guidelines by the International Office which may be applicable to your project, establish your own project guidelines as directives to be followed by your staff.
3. If you do not have a trained and experienced purchasing officer on your staff and your logistics specialist is not experienced in this area, you must plan to devote considerable time to this function while critical funds are being expended.
4. In local procurement, insist that your staff use a procedure of formal competitive bids or telephone bids when buying goods or services.
5. In the guidelines to your staff, consider establishing a ceiling on the amount of funds the staff may commit. Costs in excess of this ceiling would require either your approval or the International Office's approval.
6. If possible, include competent and trustworthy local nationals to assist in the local procurement. Using expatriate buyers will drive up the prices on the local markets.
7. Make certain your procurement staff has a copy of the World Vision Field Financial Manual and complies with these published policies from the International Office.
8. When purchasing goods, obtain the services of an independent surveyor to verify the quality of goods.
9. Insist that your staff have a system of timely pay ments for goods and services received. Make certain there is clear agreement in writing with the vendors concerning the type of currency to be used in payments.
10. When international procurement is required, it is important to have immediate access to the key suppliers. Do not hesitate to seek assistance from World Vision support offices, in or near the countries where orders are to be placed. Ensure that the users are involved in the writing of specifications before orders are placed.
11. When items are procured from funds authorized by USAID grants or by cooperative agreement, be alert to additional restrictions involving the purchases.
12. When in doubt about what proper procurement actions to take, do not hesitate to contact the International Office for guidance prior to the commitment of funds.

Transportation *(Chapter VI)*

1. When your project calls for the distribution of food commodities or other relief goods, where will you obtain your trucks?
2. Before you consider purchasing trucks, make certain that your logistics staff has carefully surveyed the availability of leased or rental trucks that are capable of doing the job.
3. If the only alternative is to purchase trucks, make absolutely certain that you first have the necessary funds in your budget and that you have the means of supporting these trucks with responsible drivers, mechanics, shop facilities and tools, spare parts, fuel, warranty maintenance, etc. Be prepared to justify the purchase of project trucks to the International Office.
4. When buying trucks, shop around to get the best and most reliable truck for the money. An important factor to consider is the most responsive dealer who can provide available maintenance support, to include warranty maintenance repairs and access to an ample supply of spare parts, tires, etc.
5. In certain countries fuel is difficult to obtain. By owning your own project trucks, you will be re-

Smiling, happy, contented faces are the result of a relief project's successful logistics operation.

sponsible to provide your own fuel. The storage of fuel could be a major problem.

6. You must arrange to provide sufficient administrative vehicles for the staff. It is advisable to purchase these vehicles since rental and leasing costs are normally very high. To simplify care and maintenance of these vehicles, stay with the same brand name for all vehicles.
7. Insist that the transport coordinator or the logistics specialist prepare a detailed distribution plan. This will become a part of the overall logistics plan. Refer to the "Transportation" chapter for formats.
8. Make certain there is a reliable system of monitoring the distribution of relief goods.
9. Keep in mind that the most costly means of distribution is by chartered aircraft. Before using this means of transport, ensure that sufficient funds are available and that the use of the aircraft is justified by the possibility of saving lives. Obtain necessary approvals from the International Office. The use of aircraft could be by air delivery or airdrop (using military crews).
10. In certain countries rail is available for large-scale transport to distribution centers. It is necessary to evaluate the reliability and security of the railroad since inefficiency and delays would make the use of rail inadvisable.

Materials Management
(Chapter IX)

1. From the standpoint of a project manager who is responsible for the operations of an emergency relief project involving the distribution of large tonnages of relief commodities, this function of logistics is extremely important and should receive considerable attention.
2. Soon after the project has been conceived and the PPF has been submitted, request that the logistics specialist verbally brief you on his plans for implementing this function. Emphasis should be placed on his plans for establishing a stock control and accountability system. Necessary forms should be available for implementation (refer to Chapter IX for an explanation of these forms). This system should be ready to implement immediately upon arrival of the first shipment of commodities.
3. Stock control means the proper accounting for all the supplies, equipment and commodities in the inventory at all levels of the project. This accountability begins when the ownership of the commodities first passes to World Vision. You can expect, and should be prepared for, any audit inspector to delve into your system very carefully. If the system is ineffective, expect your project to be formally criticized.
4. If your commodities are furnished by the U.S. government (PL480), insist that you be assigned an experienced commodity manager, or someone who is knowledgeable in USAID regulations and accounting and reporting procedures.

5. In the early logistics planning of a project, you should work closely with your logistics specialist in selecting distribution centers where you will receive the commodity shipments. These centers should be reasonably close to the depressed population, be readily accessible to long-haul trucks, and have sufficient covered storage (warehousing) to receive and store the commodities.
6. Use open storage only as a temporary expedient, not as a long-term solution for storage. The section, "Storage," in this chapter summarizes the criteria for the use of open storage.
7. Ensure that competent and trustworthy warehousemen and clerks who maintain accountability records are included in your logistics team. These personnel will probably be selected local nationals.
8. Security and fire protection are two major criteria you should inspect for in organized warehousing. Also, insist on absolute cleanliness; there should be no evidence of rodents or insects. Make certain there is ample space between each stack and the stacks and the walls of the building.
9. Often during the planning for a project, the question arises about how large must a warehouse be to hold so many tons of commodities. There is a section on "warehouse capacities" which includes simple calculations that can be used to answer such questions.
10. Appendix IV includes useful "conversion factors" that are common to logistics operations.
11. At the end of Chapter IX, there is a convenient checklist for covered storage which you should refer to before discussing warehouse operations with your logistics specialist.

Maintenance *(Chapter X)*

1. This chapter refers to the maintenance of equipment as well as the maintenance of trucks and vehicles owned by World Vision. Particularly if this capital investment is great, you should insist on the assignment of a trained and experienced master mechanic, or maintenance chief.
2. If your vehicles and equipment are properly cared for by the users and operators, this will greatly reduce (not eliminate) your maintenance requirements. This is called "preventive maintenance."
3. Insist that your logistics specialist organize and conduct a training program for drivers and mechanics, in proper driving procedures and preventive maintenance.
4. A scheduled maintenance program is a regularly planned inspection and servicing of vehicles and equipment. The establishment and enforcement of such a program will prevent untimely breakdowns--resulting in extended periods of deadlined equipment.
5. When operating your own trucks and vehicles, make certain there is a system for providing ample spare parts and tires. Also, an adequate workshop space, with the tools and equipment, should be available as well as necessary mechanics to keep the equipment running. There are added difficulties when you own and operate project vehicles, which can distract from your primary ministry of helping distressed people.
6. The availability of spare parts is critical to keeping your trucks and vehicles on the road. As a manager, inspect the system set up by the maintenance chief for replenishing these spare parts as they are consumed. The logistics specialist should be on top of the maintenance program at all times. Without reliable trucks and vehicles, the distribution cannot get done.

Other Logistics Functions *(Chapter XI)*

1. No two relief projects are the same. As for logistics functions, some will be common to all projects, but there may be additional functions that would receive more emphasis in certain operations. This handbook has emphasized the need for the logistics operations to remain sufficiently flexible to meet the support requirements of the project.
2. In the initial logistics planning for the project, this is the time to determine specifically what functions will be required and how the logistics team should be organized to support the project.
3. Refer to this chapter to identify other logistics disciplines that will be necessary for the success of your project. For example, there may be a major requirement for radio communications, not only for trucks but also from base stations to remote feeding centers. The acquisition of several radios would justify a radio technician as a valuable addition to the logistics team.
4. When your project teams are operating in remote locations, you must ensure that the logistics team has a workable emergency evacuation plan for members of the field team to meet serious crises involving sickness or injury.

Appendixes and Index

This is what it is all about.

Sample Agreement for Services

Appendix I

THIS AGREEMENT entered into this ______ day of __________ , 1986, by and between World Vision International, a California nonprofit corporation, hereinafter referred to as "WVI" on the one hand, and _________________ (name of contracting party), hereinafter referred as "Provider," on the other hand.

WHEREAS, WVI is currently engaged in providing relief assistance to the people of ______________ through the delivery of specified food commodities, based on an agreement with the United States Agency for International Development (USAID) to provide _______ tons of such food commodities (hereinafter referred to as the "Project"), and desires to engage the services of the Provider for assistance in the Project through the rendition of Provider's services on the terms and conditions hereinafter provided; (set out basis of agreement, e.g., the purpose and background to the contract).

WHEREAS, Provider desires to provide professional services to WVI as described in Exhibit A;

WHEREAS, Provider has professional expertise and competence to provide said services in a professional manner;

WHEREAS, Provider has estimated, as indicated below, the cost of providing the professional services described in Exhibit A;

NOW, THEREFORE, in consideration of the above recitals and other good and valuable consideration, it is agreed as follows:

1. Term

This Agreement shall commence on ___________ and shall continue until _______________ or until terminated as provided in Paragraph 2 below.

2. Termination

This Agreement may be terminated by either party upon ten (10) days' written notice. Prior to such termination by either party, the one requesting said termination shall notify the other in writing of the grounds for termination and the notified party shall have fourteen (14) days following notice to attempt to cure the grounds for termination. Within ten days following termination of this Agreement by either party, all vehicles, tools, equipment and other properties obtained or prepared pursuant to this Agreement shall be given to WVI.

3. Services

Provider shall provide to WVI the professional services described in Exhibit A commencing as of _________ . No additional services shall be provided without the express written consent of WVI. All services rendered to WVI by Provider shall be performed in a timely, professional, competent and non-negligent manner.

4. Responsibility of WVI

It shall be the responsibility of WVI (set out details of responsibilities, e.g.):

a. Find and supply

____

____

b. Provide directions as needed concerning the destination of the commodities and all other necessary instructions regarding ____________ the project to be carried and to act as a liaison person to maintain contact with the Provider at all times.

c. Pay _______________ on the basis of invoices and other documentation submitted as evidence of the performance of such services.

d. Provide food and lodging at ____________ as required and agreed upon in advance by the parties for which WVI shall be reimbursed in accordance with Appendix C. The provision of housing shall be dependent upon availability.

e. Pay the port charges and storage fees if required prior to loading of goods onto Provider's vehicles.

f. Assist Provider in obtaining visas, driver's licenses, travel and work permits, clearance for expatriates and other required documentation necessary to function within ____________.

5. Fees and Costs

a. Provider shall be paid a fee for work performed as set forth in the schedule attached as Exhibit B, provided, however, that Provider's fee shall in no event exceed the maximum sum of ____ dollars.

b. As an advance payment, WVI has paid Provider the sum of ___________ to enable Provider to purchase necessary tools, equipment and vehicles and employ necessary staff. This payment shall be amortized by 12 monthly deductions, starting on ____ of _______. (Optional-- do not use unless required.)

c. In the event of either WVI or Provider being unable to continue their performance under this Agreement due to circumstances outside their control, the amount of expense actually paid by Provider shall be reimbursed by WVI, plus two percent; or WVI shall pay $_______ per ton for food commodities already delivered under this Agreement, whichever is the higher of the two sums. The initial payment of _________ made by WVI to Provider shall be returned after the costs incurred prior to such termination have been deducted. (Optional-- you may want some other method of determining amount due in case of inability to perform.)

d. For all fees and costs claimed by Provider, Provider shall submit within ten days of the end of each calendar month invoices for payment based upon services furnished and costs incurred during the previous calendar month. (Payment by WVI shall be made on the balance due after first deducting the monthly payment due from Provider of ___________. (Only include if a down payment is made.)

6. Devotion of Time

The Provider shall devote such time to the performance of its duties under this Agreement as is reasonably necessary for satisfactory performance. Should WVI require additional services not included in this Agreement, the Provider shall make a reasonable effort to fit such additional services into its time schedule without decreasing the effectiveness of its performance or its duties hereunder. Additional personnel shall not be hired to complete such added services without the written consent of WVI.

7. Insurance

a. Provider shall maintain a policy of liability insurance in the minimum amount of ____ dollars per occurrence to cover any claims arising out of the performance of its services under this Agreement. The policy shall state that WVI is named as an additional insured "as WVI's interests may appear." Further, the provider shall indemnify, save harmless and defend WVI from any such claims arising from any act or omission of the Provider or its agent. The Provider shall be considered an independent contractor and not an employee of WVI under this Agreement.

b. WVI shall maintain insurance on the vehicles themselves. However, in the event of damage to a vehicle, Provider will repair the vehicle to the best of its ability. If the damage is so extensive that the vehicle is irreparable, WVI shall bear the loss of the vehicle.

8. Ownership of Materials

All tools, equipment, vehicles, supplies, maintenance base and other materials bought or produced pursuant to the terms and conditions of this Agreement are the property of WVI provided WVI shall have paid Provider as required herein. All such items will be handed over to WVI at the expiration of this Agreement.

9. Representatives of WVI

The individuals identified below, or other individuals later identified in writing by one of the individuals identified below, are designated as the sole and only representatives authorized on behalf of WVI to act in its behalf or receive information or communications from Provider.

____________________, Director of ___________,
or in his absence_________________.

10. Entire Agreement

This Agreement supersedes any and all other agreements, either oral or in writing, between the parties hereto with respect to the subject matter herein and no other agreement, statement or promise relating to the subject matter which is not contained herein shall be binding. The Agreement cannot be changed or terminated orally.

11. Assignment

Neither this Agreement nor any duties or obligations hereunder shall be assignable by the Provider without the prior written consent of WVI. In the event of an assignment by the Provider to which WVI has consented, the assignee or his legal representatives shall agree in writing with WVI to personally assume, perform and be bound by the covenants, obligations and agreements contained herein.

12. Successors and Assigns

Subject to the provision regarding assignment, this Agreement shall be binding on the heirs, executors, administrators, successors and assigns of the respective parties.

13. Arbitration

a. Unless otherwise indicated, all disputes, claims and questions regarding the rights and obligations of WVI and Provider under the terms of the contract documents are subject to arbitration. In case of dispute, WVI or Provider may make a demand for arbitration by filing such demand in writing with the other party. The demand shall be made within five (5) days after a dispute first arises.

b. If WVI and Provider agree on the selection of an arbitrator, there shall be only one arbitrator. If no agreement is reached within five (5) days after demand for arbitration, there shall be three arbitrators, one named in writing by WVI and a second by Provider within fifteen (15) days after demand for arbitration, and a third chosen by the two who are appointed. If there is one arbitrator, his decision shall be binding; if there are three arbitrators, the decision of any two of them shall be binding. No one shall act as an arbitrator who is in any way financially interested in the work or in the business affairs of either WVI or Provider.

c. Should either WVI or Provider refuse or neglect to appoint an arbitrator or to furnish the arbitrators with any necessary papers or information, they are empowered by both parties to proceed ex parte. The decision of the arbitrators shall be a condition precedent to any right of legal action that either party may have against the other.

d. The arbitrators, if they so deem, are authorized to award to the party whose contention is upheld such sums as they deem proper for the time, expense and trouble incident to the appeal and, if the appeal was taken without reasonable cause, damages for delay. The arbitrators shall fix their own compensation, unless otherwise agreed on, and shall assess the costs and charges of the arbitration on either or both parties.

e. The services to be provided shall not be interrupted or delayed during any arbitration proceeding except on written agreement by both parties.

f. Unless otherwise agreed in writing by the parties, any arbitration proceedings which take place in accordance with the provisions of this Article shall be held in Los Angeles, California, or at a place selected by the arbitrator or arbitrators, as the case may be.

14. Limitations

a. Time is of the essence and Provider agrees to ______________________ as expeditiously as possible. (In the event of a railway or any other demurrage occurring due to an unsatisfactory performance level by Provider, then Provider shall be liable for such demurrage and charges--optional.)

b. WVI shall have the right to conduct, with appropriate notice, on-the-spot checks of ________ and any related documentation pertaining to the services rendered under this Agreement, including Provider's books and records as they pertain to WVI activities prescribed in this Agreement. Provider shall maintain all books and records pertaining to the provision of services hereunder for a period of at least three (3) years from the date of expiration of this Agreement.

c. This Agreement does not provide for the transportation of personnel or other individuals under its terms. (Optional--use only if transportation agreement.)

15. Governing Law

The validity of this Agreement and of any of its terms or provisions, as well as the rights and duties of the parties hereunder, shall be governed by and construed in accordance with the laws of the state of California. (You may have to omit this if local law requires that it governs.)

16. The Incorporation

Each exhibit and document referred to in this Agreement is hereby incorporated.herein as if set forth in full.

Executed at________________ on the day and year first above written.

WORLD VISION INTERNATIONAL

By:____________________________________
Its:____________________________________

PROVIDER

____

By:____________________________________
Its:____________________________________

Exhibit A: Professional Services

(Set out the exact services to be provided. This is a sample only.)

Provider will provide professional services to WVI regarding the collection, transportation and delivery of all goods requested by WVI from point of entry into __________________ or any other location within ______________ to the projects within the country. This delivery may include destinations which are required as intermediate points.

To complete this program Provider, among other things, shall accomplish the following:

1. Maintain, repair and service a fleet of not less than___________ as outlined in Exhibit C hereto, in excellent running order. The standard of maintenance to be compiled with is that not more than ten percent of the minimum number of the fleet shall be off-road at any one time, due to maintenance or repair difficulties. Thus 90 percent of the minimum operating fleet will be in action at any one time.

2. Maintain, repair and service all other "small" vehicles belonging to WVI. The same standard of maintenance as noted in No. 1 above shall apply to the maintenance and service of these small vehicles. Small vehicles shall include landcruisers and other administrative vehicles.

3. Transport all goods required by WVI to be transferred from one location to another.

4. Train______________ nationals to be able to take over the project within 12 months of the commencement of this Agreement.

5. Assume responsibility for security of all goods, whether at the port and/or in Provider's care in storage prior to being transported, as well as for the period during which the goods are actually being transported.

6. To transport commodities and other goods for other NGOs as opportunity arises as long as WVI's goods are not delayed. Prior to agreeing to transport such items, Provider shall obtain the written permission of WVI which permission shall include an agreed upon price. Provider shall pay to WVI 50 percent of the amount charged to the NGO. This may be subject to audit by WVI.

Exhibit B: Fee for Professional Services

(Set out the maximum amount due under the contract, and the conditions and method of payment.)

WVI agrees to pay Provider at the rate of $_______ (U.S.), per metric ton for all commodities transported from ____________________to their initial destination up to ___________ tons. (After___________ tons of commodities have been carried at rate payable will be $_____ (U.S.) per ton--optional.) No additional payment will be paid for the subsequent transport of the same commodity.

When items other than commodities are being transported, WVI agrees to pay ______ per day per vehicle for jobs requiring more than three (3) days to complete. Where the transport is completed in less than three (3) days, no additional compensation outside of actual field costs will be paid.

All items purchased under this Agreement shall revert to WVI upon termination of the agreement as per Section 8.

The rate per metric ton includes, but is not limited to, the following:

1. Salaries, wages, housing, air fare, benefits and other expenses incurred for up to______ expatriate staff and all national staff.

2. Communications expense.

3. Spare parts for trucks, trailers and small vehicles.

4. Tires and oil for all vehicles.

5. Fuel for trucks (excluding small vehicles).

6. Supplies, tools and other materials including a computer, necessary equipment for a mobile maintenance unit should it be required, air compressor, impact wrenches, spray gun, other special tools, paint and steam washer.

7.______ small vehicles for use by Provider.

8. Liability insurance for Provider and its agents and other insurance required by the government of _____________, except for insurance to cover the vehicles involved in the transport program which shall be borne by WVI.

WVI Policy on Execution of Contracts

Appendix II

Board of Directors' POLICY MANUAL	POLICY TITLE **Execution of Contracts**	DATE ORIGINIATED **7/18/84**	DOC # **3350**
	SECTION TITLE **Legal**	DATE REVISED	PAGE **1 of 1**

Summary All contracts and agreements entered into between World Vision International and any outside party must be executed by the individual designated to exercise such authority.

Policy It is the policy of WVI to restrict the number of individuals who may execute contracts on behalf of WVI, or otherwise legally obligate the corporation. A designated signee must execute any agreement or contract on behalf of WVI unless one of the following conditions exist:

> The contract or agreement is a renewal of an existing lease or other agreement not longer than one year in duration, the terms of which are identical to the original lease, contract or agreement.
>
> The only governing terms of the contract or agreement are the terms and conditions set forth in the purchase order of WVI.
>
> The agreement calls for the provision of services which are rendered as needed and are paid automatically upon the submission of an invoice pursuant to a written agreement or letter of understanding previously executed by a designated signee.

For all other contracts, agreements, leases and any other document obligating WVI, a written agreement is required, and it must be executed by one of the following:

The President of WVI; or

The Executive Vice President of WVI; or

The VPAF of WVI; or

The individual authorized in writing to be the designee of one of the above officers of the corporation.

APPROVAL ***DATE***

Issuing Vice President

Executive Vice President/President 3 October 1984

Position Description: Logistics Specialist

Appendix III

Date Prepared: Revised - October 1985

Purpose of Action

To participate as a member of Project Management teams of World Vision International, an evangelical Christian humanitarian organization, in providing emergency relief aid to drought-stricken countries of Africa.

Major Responsibilities

- Develop concepts and plans for the logistics support of emergency relief projects in designated drought-stricken countries in Africa as a member of Project Management teams.
- Evaluate in-country logistics capabilities to support relief projects and identify any logistics constraints.
- Major logistics functions include:
  -- Transportation, all modes
  -- Maintenance (vehicles and equipment)
  -- Procurement
  -- Materials management, to include stock control, warehousing and accountability.
- Be prepared to implement logistics plans and direct logistics operations involving receipt, storage, accountability, and arranging transshipment of relief goods to end users.
- On relief projects where World Vision-owned vehicles are assigned, be prepared to organize and direct control and maintenance of these vehicles and equipment.
- As required, be prepared to assist in the logistics training of local nationals.
- Manage local purchasing offices and supervise local buyers.
- Be prepared to supervise logistics specialists assigned to subordinate relief projects.
- Prepare written logistics progress reports to higher headquarters.

Knowledge, Skills and Abilities

- Must have at least eight years' experience in logistics assignments as a manager or as a hands-on operator involving specific functions previously listed. It is recognized that expertise may be greater in some functions than in others. Basic knowledge of all functions is required. Transportation management skills are particularly desirable.
- Must be a good team player to work closely and cooperate with other World Vision project members.
- Must be able to employ good tact and diplomacy in relationships with U.S. and African government officials.
- Must have basic knowledge of developing contract provisions for logistics services.
- Must be flexible and able to adapt to a variety of living and working conditions.
- Must have good communication skills, both oral and written.

Duration

- Position is generally by contract arrangement with World Vision International for a minimum period of one (1) year. Position could be a permanent hire.
- Overseas assignments will normally be in drought-affected countries of Africa.
- Salary will be based on the extent of experience and qualifications of the applicant and position assigned.

Conversion Factors

Appendix IV

To Convert From	*To*	*Multiply By*
Length		
Yards (1 = 3 ft = 36 inches)	Meters	0.9144
Meters (1 = 100 cm)	Yards	1.0936
Miles (1 = 1,760 yards)	Kilometers	1.609
Kilometers (1 = 1,000 m)	Miles	0.6214
Meters	Feet	3.281
Feet	Meters	0.3048
Area		
Yards^2 (1 = 9 ft^2)	Meters^2	0.836
Meters^2 (1 = 10,000 cm^2)	Yards^2	1.196
Acres (1 = 4,840 yd^2)	Hectares	0.405
Hectares	Acres	2.471
Miles^2 (1 = 640 acres)	Kilometers^2	2.590
Kilometers^2	Miles^2	0.386
Meter^2	Feet^2	10.76
Feet^2	Meters^2	0.0929
Volume		
US gallons	UK gallons	0.8327
UK gallons	US gallons	1.2009
US pints	Liters	0.473
Liters	US pints	2.113
US gallons (1 = 8 pints)	Liters	3.785
Tons (shipping)	Feet^3	40.00
US gallons	Feet^3	0.1337
Meters^3 (1 = 1,000 liters)	Yards^3	1.308
Yards^3 (1 = 27 feet^3)	Meters^3	0.765
Meters^3	Feet^3	35.31
Feet^3	Meters^3	0.02832
Weight		
Ounces (oz)	Grams (g)	28.35
Grams	Ounces	0.0353

Weight *(continued)*

Pounds (1 = 16 oz)	Kilos	0.4536
Kilo (1 = 1,000 g)	Pounds	2.205
US short tons (1 = 2,000 lb)	Metric tons	0.907
US long tons (1 = 2,240 lb)	Metric tons	1.016
Metric tons (1 = 1,000 kg)	US short tons	1.102
Long tons	Short tons	1.120
Short tons	Long tons	0.8929

Temperature

Celsius = 5/9 (Fahrenheit - 32)

Fahrenheit = 9/5 C + 32

Weight of Water

1 liter = 1 kg; 1 US gal = 8.33 lb; 1 UK gal = 10 lb

Sample Calculations

(1) Weight of a 50 kg bag in pounds? *50 x 2.205 = 110.25 pounds*

(2) How many 50 kg bags is 200 MT? *200 MT x 1,000 kg = (200,000 lbs X 2.205)/110.25 = 4,000 bags*

(3) How many acres in 12,000 hectares? *12,000 hectares x 2.471 = 29,652 acres*

(4) How many feet^2 in 5,000 meters^2? *5,000 meters^2 x 10.76 = 53,800 feet^2*

(5) How many pounds does 15 US gallons of water weigh? *15 US gals x 8.33 = 124.95 pounds*

(6) How many cubic meters are there in 2,762 cubic feet? *2,762 feet^3 x 0.02832 = 78.22 meters^3*

(7) How many miles are there in 1,250 kilometers? *1,250 km x 0.6214 = 776.75 miles*

(8) What is the temperature in Fahrenheit when it is 40 C? *9/5 x 40 + 32 =104 F*

Sample Outline of a Logistics Plan

Appendix V

Subject: Logistics Plan
Date:
Identification of Project:
Location:
Participating Partners:

I. General Concept (Chapters I and II)

This paragraph provides for a brief narrative description of how the project will be logistically supported. If the relief project involves distribution of commodities, include how and where the World Vision field team will receive, store and distribute these commodities.

A general description of the overall logistics concept is an important part of the plan. In fact, this provides a good lead into a breakdown of each function that follows. If this paragraph could be included in the Project Proposal Form (PPF) that is submitted to the International Office for project approval, it would be helpful to the management reviewing the PPF. It provides assurance to management that the logistics aspects of project support have been considered. Any major logistics constraints should be included in this paragraph. It would be appropriate here to include the major resources, such as capital equipment, that will be needed to implement the project.

II. Organization (Chapter III)

A simple organizational chart would be sufficient for this paragraph. This will identify the major logistics functions involved in this particular project. It would be appropriate here to clarify any unusual reporting channels for certain sections. For example, if there is an aviation section, the project manager may specify that the logistics organization provide the administrative support for the section, but he may select himself to maintain direct control of the aircraft or of any air operations.

The organization chart should include the total number of people required for each section, specifying what positions could be filled by local nationals or if they should be expatriates.

III. Logistics Personnel (Chapter VII)

This paragraph should elaborate on the personnel requirements covered in the previous paragraph. Any specialized requirements should be singled out here. For example, if the project will include a large radio communications requirement, a radio specialist should be requisitioned. If the project plans to purchase vehicles, trucks and other capital equipment, a transportation coordinator and a master mechanic should be requisitioned.

Once the project manager has approved the logistics plan, immediate personnel action can be taken by initiating appropriate requisitions. In the event your project will be handling commodities furnished by the U.S. government (PL480), the project will require a commodity manager, or others knowledgeable in the administration and reporting in accordance with pertinent USAID regulations.

IV. Coordination (Chapter IV)

List and briefly explain the extent of coordination required with other agencies. This would include local government agencies, in-country USAID officials and other Private Volunteer Organizations (PVOs). In cases where a partnership arrangement has been made with other PVOs, an explanation would be appropriate of exactly what contribution the partner(s) will make to the project.

V. Procurement (Chapter V)

- What relief supplies can be purchased locally?
- Is there a competent, trustworthy local national on the staff to do the local buying?
- Identify likely sources for international purchases.
- Are copies of World Vision's Field Financial Manual and other published policies and procedures available for reference by the procurement staff?
- Will there be a requirement to solicit formal written bid offers for any goods and services to be used on the project?
- Is it expected that any formal contracts be required

in obtaining goods and services? If so, any contracts must be cleared prior to execution with the International Office legal staff.

- Has a timely method of payment to the vendors been established for goods and services specifying type of currency to be used?
- For international purchases, what method of transport will be used to deliver the goods and how will these transport costs be funded?
- For local purchases, is an independent surveyor required to verify the quality of goods?
- Prior to solicitation of bids, has a clear and concise specification been prepared?
- What approvals are required prior to placing orders for any costly capital equipment?
- If you are using funds authorized by USAID grants, what are the additional restrictions in the use of these funds?

VI. Transportation (Chapter VI)

- If the project involves the distribution of commodities, what is the source of trucks?
- If the decision is to purchase World Vision trucks, can this be justified with the International Office?
- When requesting authority to purchase trucks, include plans to support the ownership of these trucks with mechanics, maintenance shops, spare parts and tires, fuel, drivers, liability insurance, etc.
- Prepare a distribution plan (refer to Chapter VI).
- Where will the ownership of the commodities pass to your field team?
- Will bills of lading be used? If so, what are the plans for controlling them?
- When contracting for commercial transporters, have competitive bids been solicited?
- When using commercial transporters, has a system been established for controlling cargo?
- Are there any other modes of transport available to be used? Briefly describe their capabilities. (Chapter VII)
- If any chartered air operations are anticipated, has permission been requested from the International Office? Can these air operations be justified?

VII. Stock Control and Accountability (Chapter IX)

- Has an effective, workable system of accountability of commodities been established? Briefly describe the system.
- Are appropriate forms available in ample quantities?
- Are the storage facilities and personnel available to receive the commodities?
- Are there several camps or distribution centers to warrant a centralized stock control system?
- Has a periodic scheduling of physical inventories been set up for all storage and depot locations?

VIII. Storage (Chapter IX)

- Briefly describe the storage facilities that will be available during the project.
- Will any open storage be required; if so, for how long?
- What system for the security and fire protection of all storage facilities has been provided?
- How is the warehouse complex to be organized when handling more than one commodity?
- Has a frequent inspection schedule been planned for the warehouses?
- What insect control measures have been planned for the warehouses?
- Have you arranged for adequate warehousing to handle commodities?
- Is there a labor force available to handle the commodities in and out of the warehouses?
- Are the warehouse locations readily accessible to long-haul trucks?
- Are the warehouses dry and well-ventilated?

IX. Maintenance (Chapter X)

- Has there been a preventive maintenance program established? Briefly describe.
- In order to maintain World Vision-owned trucks, vehicles and equipment, are adequate shop facilities available?
- Are there competent mechanics to maintain the equipment?
- Are there tool sets and shop equipment available?
- What is the system for ordering, replenishing and accounting for spare parts?
- Are there plans for a drivers' training program?
- What plans have been made for the storage and dispensing of fuel?
- Are there dealer facilities available to provide warranty maintenance on new equipment?

X. Other Logistics Functions (Chapter XI)

Describe any other logistics functions that may be applicable to the project.

XI. Reports (Chapter III)

Describe a plan and outline for providing periodic reports to management on the status and progress of logistics activities during implementation of the project.

Index